Beautiful Times
Luxury Soft Decoration Ten Style (Vol. 1)

漂亮时光
豪宅软装十风格（上）

中国·武汉

Elegant Baroque 优雅巴洛克

Elegant Baroque · the Combination of Chinese and Western	008	雍容华贵巴洛克 洋洋洒洒荟中西
Heaven Botto Fino	028	天堂波托菲诺
Under the Blue Sky on the Falling Silver Spring	040	上穷碧落下银泉
Facing the Sea in Spring	052	面朝大海 春暖花开
Waltz Variations	062	变奏华尔兹

French Romantic 浪漫法式

Elegant Peacock Blue	074	优雅孔雀蓝
Aesthetic Blues Stone	082	唯美蓝调
The Peach Garden	094	桃花源
French feelings on the Shore of the Fuchun River	100	富春江岸边的法式情怀
French Style in Provence	112	普罗旺斯法式风情
Interpretation of Modern Court Art	118	诠释现代宫廷艺术
French Life Aesthetics	128	法式生活美学
MTR Song Great Carpenter Built	148	港铁天颂 大匠所筑
Years of Quiet Good, Mo Flowers	158	岁月静好 陌上花开

漂亮时光 在岁月的深处编织永恒

西班牙新古典 New Classical Style of Spain

闲	168	Leisure
托斯卡纳艳阳	188	Tuscany Sunny
爱在深秋的稻谷"金"	198	Love in the Autumn "Golden Rice"
高贵与浪漫的异国风情	208	Noble and Romantic Exotic

地中海度假 Mediterranean Resort Style

巴塞罗那的记忆	220	The Memory of Barcelona
情迷地中海	238	Discovering Mediterranean
蓝白迷情	256	Blue and White Rush
理想的生活 静静的靠近	262	The Ideal Life, Quiet Near
海的声音	270	The Sound of the Sea

复古雅皮 Retro Yuppie Style

复古的青春旋律	220	Retro Youth Melody
复古的青春旋律	238	Retro Youth Melody
绅士简约主义	256	The Simplicity of Gentleman
爱的守护	262	Guardian of Love
定义奢华新概念	270	Define the New Concept of Luxury

优雅巴洛克
Elegant Baroque

雍容华贵巴洛克 洋洋洒洒荟中西
Elegant Baroque · the Combination of Chinese and Western

项目名称：万科·五龙山别墅大独栋样板间
设计公司：深圳创域设计有限公司 / 殷艳明设计顾问有限公司
设计师：殷艳明、张书
项目地点：四川省成都市
项目面积：660 ㎡
主要材料：石材、壁纸、拼花木地板、木饰面、拼图马赛克
开发商：万科地产

Project Name: Vanke · Five Longshan Villa Large Single Show Flat
Design Company: Shenzhen Creative Design Co., Ltd. / Yin Yanming Design Consultant Co., Ltd.
Designers: Yin Yanming, Zhang Shu
Project Location: Chengdu City, Sichuan Province
Project Area: 660 ㎡
Main Materials: Stone, Wallpaper, Wood Flooring, Parquet Veneer, Mosaic
Developer: Vanke Real Estate

本案强调软硬装一体化的设计理念，通过空间与软装陈设的设计语言解读巴洛克风格的特质与亮点。由空间解构开始，这是本案立意的一个高度，任何风格只有在适合的空间中绽放，才能彰显其精神价值与独特的个性。我们突破以往传统的设计定位，从平面布局开始便把巴洛克风格的奢华、恢宏与浪漫主义色彩的灵动赋予空间以强烈的立体感。从概念色彩体系、灯光体系、到艺术文化解读和造型体系几个方面入手，把十七世纪末巴洛克风格盛行时期的雍容华贵与同时期在中国传教的意大利画家郎世宁的宫廷绘画艺术相结合，巧妙而自然的中西结合，给予作品生动的依托。西方设计界流传着一个观点："无中不贵气"。十七世纪那场跨越时空的皇家宫廷艺术之花，通过中西合璧的巴洛克恋曲在空间中缓缓盛放，既是对历史的感怀，也是向巴洛克艺术风格的致敬。

位于一层的客厅会客区以宝石蓝色系为主，在西方宗教传统中宝石蓝是皇族与贵族钟爱的颜色，迷人而优雅，处处彰显贵族的气息。壁炉背后的黑金手绘壁纸，展现出主人对欧洲文化深厚底蕴的留恋与玩味。整体设计方正大气，沙发群组与壁炉、吊灯及挂饰相映生辉，天花形态曲直相生，图案与光影交汇，展示了巴洛克风格动态中的平衡美感，古琴的设置让空间在精神层面上有了更高品味的追求。

女士餐茶区以酒红色为主，区域的设置突显了人性化的关怀和优雅的生活方式，孔雀蓝与羽毛让闲雅的空间有了鲜活的气息。

二层是主卧，空间内的所有家具与软装配饰格调相同，地面古典图案的咖啡色地毯与拼花木地板、金色雕花的屏风，柔化了居室的硬朗质感，在整体营造奢华氛围的同时，床头墙上的绢上绘画《百骏图》更突显低调中的奢华。

浴室的设计紧致、优雅。孔雀摆件姿态怡然，绚丽夺目。洗手台采用烟玉大理石，搭配古典贝壳马赛克的铺砌，定制的金色椭圆镜点缀，营造出一个浪漫、舒适的空间氛围。这不是一个传统意义上的浴室，而是一个可以品酒、愉悦身心的休憩所在。

中庭空间承上启下，水晶吊灯与绽放的花形、地面流动的圆形图案都在优雅中传递出空间的气质，形与意、态与势展露出瑰丽奢华的贵族风范。

豪宅软装十风格·优雅巴洛克

地下一层大厅改设为宴会厅，这是具有仪式感的一个重要空间，宴会厅空间布局与陈设热烈、激情而又华丽，巴洛克激情艺术的气氛展露无遗，金色与蓝色的碰撞给人强烈的视觉震撼。

男孩房以海蓝色为主，同时在局部增添了白色，运动、时尚，色彩沉静，高贵中展示出年轻人的气质与修养。

女孩房粉紫色的主色调让人觉得甜美温馨，跃层空间的儿童活动区有浓厚的童话气息，缤纷的色彩洋洋洒洒造就了一片童心世界的意境。

父母房色调沉着、层次丰富，中式纹样图案与配饰青花瓷相得益彰，在光影中传递出美好的感性体验。

多功能厅色彩浓烈，打破理性的宁静与和谐，强调激情的艺术，家具、灯具、饰品、布艺……它们和着空间的韵律节奏、功能主题的转换，具有浓郁的浪漫主义色彩。

本案立意，贵在完整贯彻了软、硬装一体化的设计初衷，细细研磨，终有回响。准确的定位与理念是设计落地的根本，一个成熟的设计作品必是贴近生活又融入历史的，真正打动人的不是繁复与冗繁的造型、线条，也不是流金溢彩的水晶灯饰与装饰，而是对设计的追求与执着。沉醉奢华而又雍容尔雅，华丽只是表象，空间的情感才能真挚动人。

This case emphasizes the design concept of hardware and software equipment integration, and through the space and furnishing design language to interpret the characteristics and highlights of the baroque style, by space deconstruction start, this is the conception in the case of a highly, any style only in the space that suits bloom, Chang to show its spiritual value and unique personality, we break through traditional villa positioning, with a full suite of space, and the use of space height adding floors between the thermocline space two pronged, from graphic layout at the beginning of the baroque style luxury, magnificent and romantic color smart gives space gives the strong sense of three-dimensional. From the concept of color system, lighting system, to art and cultural interpretation and modeling system of combine the court painting at the end of the 17th century baroque style prevalent period of elegant and at the same time of missionary in China, Italian painter Giuseppe Castiglione, clever and self contingent of Chinese and western, to the works given vivid relying on. The spread of a point of view in western design field of: "no not gentry, the seventeenth century the space-time flower of the Royal Palace of art, through the combination of Chinese and Western Baroque Sonata in space slowly blooming across, is's recollections of history and to the baroque style salute.

In the living room to receive visitors area with precious stones blue department mainly, in the Western religious traditions sapphire blue is the color of royalty and nobility affection, charming and elegant, always highlight the noble atmosphere. Black painted fireplace behind the wallpaper, showing and Pondering on European culture with profound master. Overall design of the founder of the atmosphere, sofa group with a fireplace, ceiling lamps and ornaments Xiangying Shenghui, smallpox morphology of the merits of the mutual promotion of the five elements, the pattern of light and shade and intersection, showing the baroque style dynamic balance of beauty, guqin settings let a space in spiritual pursuit of higher taste.

Ms. meal tea area with wine red, regional settings highlight the humane care and elegant way of life distinguished and peacock blue and feather let elegant space with fresh breath.

On the second floor is the master bedroom, within the space of all furniture and soft decoration style is the same, ground classical pattern of coffee colored carpets and parquet wood floor, golden carved screen, soften the room tough texture, in the overall building luxurious atmosphere at the same time, bedside silk painting the Batterymarch figure "highlight the low-key luxury.

Compact and elegant bathroom design. The peacock ornaments attitude contented, gorgeous. Washing hands with smoke jade marble, the classical shell mosaic paving collocation, custom golden oval mirror embellishment, creating a a romantic and comfortable atmosphere, this is not a traditional bathroom, but can wine tasting, where the rest of the physical and mental pleasure.

Atrium space is a connecting link between the preceding and the following, crystal chandeliers and blooming flower shape and surface flow in a circular pattern are in the elegant transfer space temperament and wonderful, form and meaning, state and potential development exposed magnificent luxurious aristocratic demeanor.

Negative on the first floor of the hall is set for the banquet hall is with a sense of ritual is an important space, the banquet hall space layout and furnishings warm, passionate and magnificent, the atmosphere of Baroque art passion revealed without involuntary discharge of urine, gold and blue collision give people a strong visual shock.

Boys room with navy blue, and in the local add white, sports, fashion, color quiet and noble in the penetration of young people's temperament and self-cultivation, to bring a creative fashion, dynamic.

Girl real pink purple main colors make people feel warm and sweet, space thermocline children's activity area open and strong flavor of the fairy tale, the colors eloquent created a piece of the innocence of the world artistic conception.

Parents room tone calm, rich in level, Chinese style pattern and accessories blue and white porcelain complement each other, in the light of the transfer of a beautiful emotional experience.

Multi - functional hall space color strong, breaking the quiet harmony of the rational, emphasizing the passion of art, furniture, lamps, ornaments, fabric…… They and the rhythm of space, the function of the theme of the conversion, with a strong romantic color.

This case purposive non new, expensive complete and implement the hardware and software installed one of the original design, finely ground and eventually echo; accurate positioning and philosophy is the fundamental design landing, a mature design works must be close to the daily life history, really touched people than complicated and cumbersome shapes, lines, also not liquidity sparkling crystal lighting and decoration, but to design and persistent pursuit. Indulge in luxury and gorgeous only appearance, have an easy manner, emotion sincere and moving to space.

豪宅软装十风格·优雅巴洛克

漂亮时光

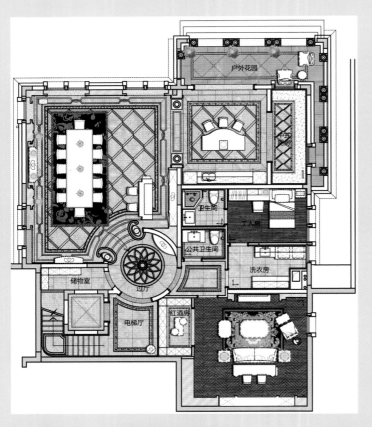

负一层平面布置图

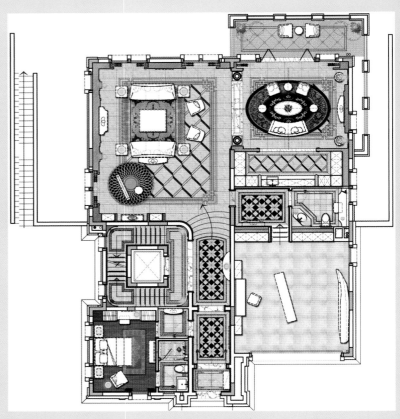

一层平面布置图

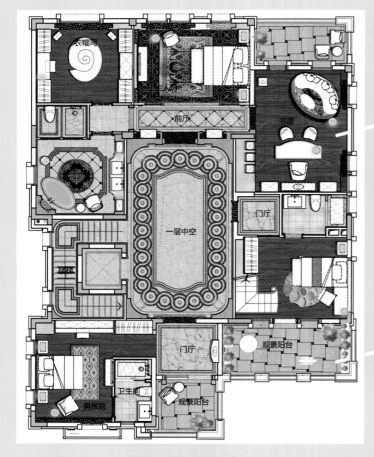

二层平面布置图

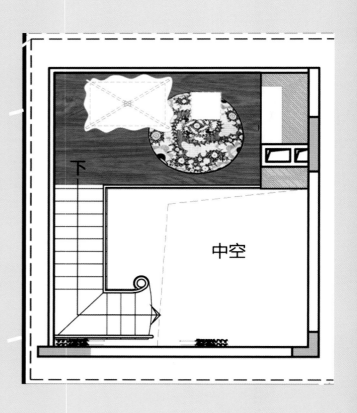

顶面布置图

天堂波托菲诺
Heaven Botto Fino

项目名称：西安卢卡小镇联排别墅样板房
设计公司：上海无相室内设计工程有限公司
空间设计师：王兵、谢萍
软装设计师：李倩
项目地点：陕西省西咸新区
项目面积：310 ㎡
摄影师：三像摄 张静

Project Name: Xi'an Luca Town of the Platoon Villa Show Flat
Design Company: Shanghai ANIMITTA Interior Design Engineering co., Ltd.
Space Designers: Wang Bing, Xie Ping
Soft Decoration Designer: Li Qian
Project Location: West Ham District, Shanxi Province
Project Area: 310 ㎡
Photography: Threeimages Zhang Jing

豪宅软装十风格·优雅巴洛克

豪宅软装十风格·优雅巴洛克

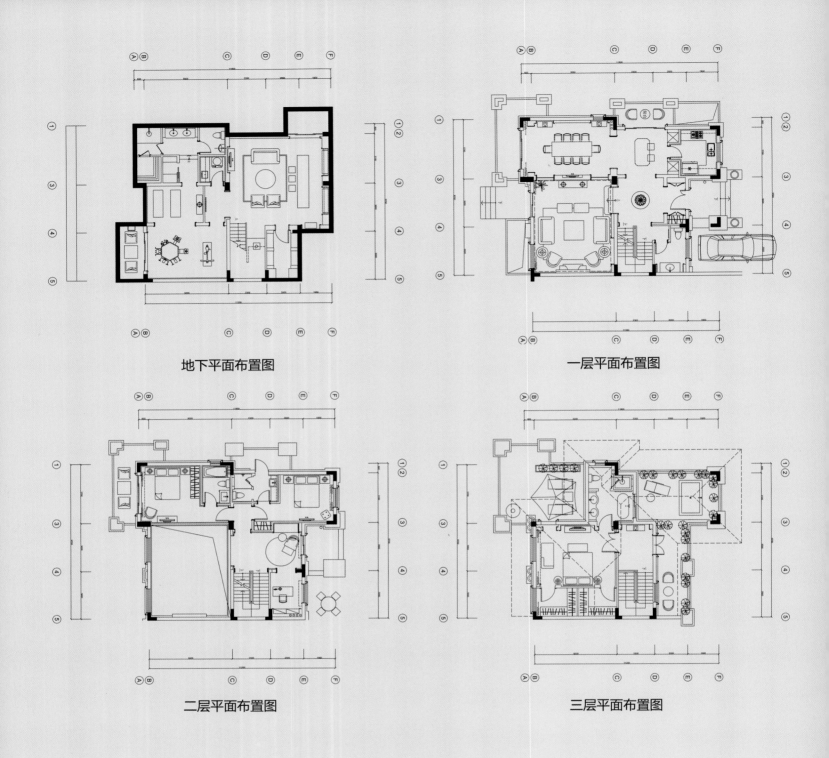

地下平面布置图

一层平面布置图

二层平面布置图

三层平面布置图

整个设计风格传承了意大利丰富的艺术底蕴，同时又融入了开放、创新的设计思想，将人们带到举世闻名的旅游天堂波托菲诺。从整体到局部，从繁杂到简单，精雕细琢，都给人一丝不苟的印象。

在本案中，我们可以看到空间形态是遵循章法的，新古典主义风格的墙面、地面的细部处理也中规中距，一方面保留了材质、色彩的大气风范，仍然可以很强烈地感受传统的历史痕迹与浑厚的文化底蕴，同时又摒弃了过于复杂的肌理和装饰，简化了线条。

为了呈现更为年轻化的空间形态，使其生活方式更具有活力，空间要求通透、自然、舒适、方便，因此我们在一层空间中将门厅、客厅、餐厅与西厨房打通，融为一体，只是将对称的门洞作为空间分割的界限。

在地下室这个空间中，将起居室与酒吧合二为一，形成一个更欢乐的娱乐空间，而相对独立又有关联的健身房、瑜伽室及水疗区又为时尚健康的生活增加了一个亮点。

在明丽的阳光里，轻声交谈，开怀大笑，尽情享受着醇香的咖啡、葡萄酒和地中海美食。每个人的脸上都洋溢着轻松、安宁、满足和幸福的表情，空气里也弥漫着香甜轻松气息。这便是许多人梦想中的天堂！

The whole design style heritage of the Italian culture rich artistic heritage, at the same time into the open and innovative design ideas, to bring people to the world-famous tourist paradise Portofino. From local to the whole, crafted from complex to simple, gives the impression, be strict in one's demands.

In this case, we can see the space form composition follow, neoclassical style wall surface detail processing also regulations in the distance, on the one hand retained the color material, the atmospheric style, you can still feel very strongly that traces the history of traditional culture and the vigorous, at the same time abandon the overly complex texture and decoration, simplified lines.

In order to present more younger configuration, the life style and form also tend to be more modern, so the space requirements of transparent, natural, comfortable, convenient, so we in a layer of space will foyer, living room, dining room and kitchen open, merges into one organic whole, only through the doorway, symmetry as a spatial segmentation defined.

In the basement space, will play room and a bar combined, forming a more joyful entertainment space, and relatively independent and association of the gym, yoga room and spa area for healthy and fashionable life increased by a central bright spot.

In the bright sunshine, talking, laughing, enjoying the fragrance of Coffee, Wine and Mediterranean Food. Each person's face is filled with light, peace, contentment and happiness, the air is filled with a sweet breath. This is the paradise of many people's dream!

漂亮时光

豪宅软装十风格·优雅巴洛克

上穷碧落下银泉
Under the Blue Sky on the Falling Silver Spring

作品名称：GI10住宅案
设计公司：台北玄武设计
设计师：黄书恒、欧阳毅、陈佑如、张铧文
软装布置：胡春惠、张禾蒂、沈颖
主要材料：银狐、黑白根大理石、镜面不锈钢、黑蕾丝木皮、银箔、金箔、进口拼花马赛克、黑白色钢烤
摄影师：赵志程
撰文：程歆淳

Project Name: GI10 Residential Case
Design Company: Taipei Basaltic Design
Designers: Huang Shuheng, Ouyang Yi, Chen Youru, Zhang Huawen
Soft Decoration Arrangement: Hu Chunhui, Zhang Hedi, Shen Ying
Main Materials: Silver Fox, Black and White Roots, Mirror Stainless Steel , Black Lace Veneer, Silver, Import Mosaic Pattern, Black and White Color Steel Baking
Photographer: Zhao Zhicheng
Author: Cheng Xinchun

豪宅软装十风格·优雅巴洛克

本案是坐落于城市新区的宅邸，有半山坡的绿意相伴，从客厅落地窗放眼望去，广场的辽阔视野，也成为居所的重要亮点，作为退休生活的启始，业主必然需要一番缜密而细腻的规划。玄武设计考虑屋主姐弟与母亲同住的实用需求，以及居住者对于美学风格的爱好，力求艺术生活化，生活艺术化，最终择以现代巴洛克为基底，以其独有的收敛与狂放，配合玄武擅长的"中西混搭—冲突"美学，铺陈空间每处轴线。

尚未进入玄关，已看到一座当代艺术作品灵动而立，既巧妙掩饰了半弧形的缺角，又以生动的童稚神情，为居所埋入活跃的生机；右边，高耸的柱式与圆形的顶盖，使人们的视觉猛然挑高，心情随之豁然开朗。经典的黑白纯色打底、定制的家具，配合景泰蓝珐琅与定做琉璃，东西文化的灵活互动，为访客带来第二重震撼。

屋主因业务所需，时有社交与公务的需求，特别需要一个大气却又有趣的客厅空间，活络人际交往。是故，玄武设计着重天然风光与人为艺术的调和，保留大型落地窗与沙发的间距，后者特别选用进口原版设计，呈现简练利落的现代风情。与此反之，中央大胆置入以艳紫、宝蓝与金黄三者交织而成的地毯，强化了简约与繁复的冲突美感，亦展露出皇室家居的大度。

豪宅软装十风格·优雅巴洛克

抬眼向上，一盏华丽的银色花朵吊灯灿烂夺目，使人倍感震撼。这座取材自苗族银饰的大型艺术品，为玄武设计与当代艺术家席时斌先生共同创作，外围借用鸢尾花意象，曲折的花饰包覆着核心，间隙镶嵌彩色琉璃，使打底的银灰色更显时尚，每当开关按下，艺术品外围即有五彩灯光流转，可因不同情境而切换，上缀羽饰的大型银器环绕着核心缓缓移动，隐喻着天文学——恒星与行星的概念，呈现着自然与人文的灵动对话。

穿越廊道，可进入屋主的阅读空间。两处各以深、浅为底，再各自于细微处呈现相反的色彩诠释。诸如，主卧书房延续着公共空间的半圆形语汇，引导访客进入皮质沙发、深色书柜、石材拼花共构的豪气场域，另外，这里使用清淡色泽的织毯，大幅提升空间的律动感；第二主卧的书房，则纳入半户外的开阔设计，以白色底板铺底，却照样使用黑色书柜与铁灰沙发，抢眼的小号造型灯具，具体而细微地体现了屋主的喜好，展现内外呼应的生活态度。

因应屋主对于公私界线的看重，玄武设计亦将此概念纳入考虑，公共区域的门扉使用白色，给人亲近、纯净之感；进入私区则以黑色区隔，带有隔绝、凸显"正式"的意义。进入次要空间，棋牌室与餐厅分据左右，二者均以白色为主调，黑白格地板，置入经典款水晶灯，搭配巴洛克花纹座椅、鸽灰抱枕，远观近看，各有深韵。

为使主、客卧有舒适的感受，卧房采用一贯的轻柔色泽，再以方向不同的线条勾勒空间表情，诸如主卧简练的长形线板，与金黄床褥、浅蓝地毯相映成趣，减少过度堆砌的冗赘感；其余卧房则以湖水绿、天空蓝为点缀，在纯白、浅灰的基调里，窗帘、床褥与地毯稍有呈现，与牡丹纹床背板的繁复，共谱出屋主悠闲淡雅的生活情趣。

This case is located in the urban district of residence, accompanied by the hillside and a half green, from the living room window looking ahead, square of the vast field of vision, has become home to highlight important, as retirement startup will inevitably need to some careful and detailed planning. Xuanwu design consider homeowners siblings and mother live with practical demand, and occupants taste for aesthetic style, and strive to "art of life" art of living, ultimately choose to modern Baroque substrate, with its unique convergence and wild, with basaltic good mix of Chinese and Western conflict aesthetics, elaborate space every inch of the axis.

Have not yet entered the entrance, has seen a contemporary works of art smart standing is a cleverly disguised semi arc missing angle and lively childlike look, is home to the buried active vitality; into the right, starting towering column and circular top cover, vision suddenly choose high, the heart suddenly see the light, the classic black and white solid grounding, mid grilling custom furniture, with cloisonne enamel and custom glazed, flexible interaction of eastern and Western cultures, for visitors to bring the double shock.

Don't in the vestibule of the pure color, to house main business, friendship and business demand and, in particular, to a device and interesting living room space, active interpersonal dialogue. Therefore, Xuanwu design focuses on the natural scenery and human art reconcile, retain large window and sofa spacing. The latter is especially used imports of the original design, showing a neat and concise modern style. And conversely, central bold placement by the brilliant violet, blue and golden yellow three interwoven into the carpet, strengthen the simplicity and complexity of the conflict of beauty, also stemming the flow of the French royal court although magnanimous.

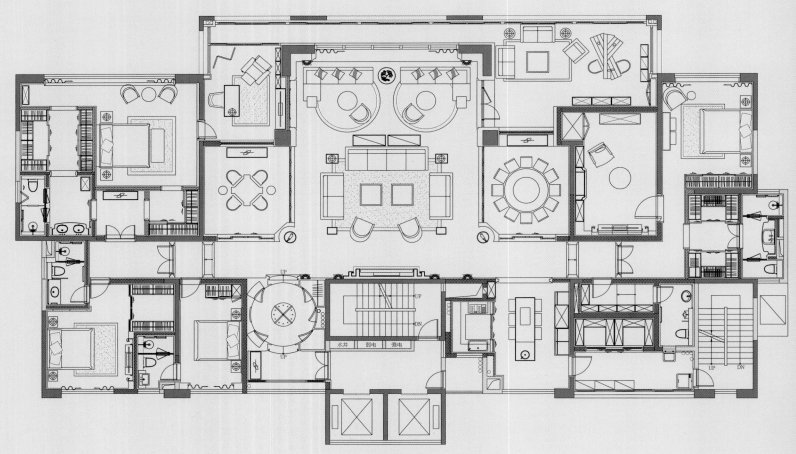

平面布置图

Lift an eye to, a magnificent silver flowers brilliant, make the person feel shocked, the drawn from Miao silverware of the large art, Xuanwu design and contemporary artists Xi Shibin Mr. co creation, external borrowing iris flower image, the twists and turns of the floral coated core, gap mosaic color glaze, the backing of the silver obviously fashion, whenever a switch is pressed, periphery of the art is colorful lighting circulation, because of the different context switching, compose the plume of large silver surrounds the core moves slowly, a metaphor: stars and planetary astronomy concept, showing a natural and human clever dialogue.

Through the corridor, the owner can enter the reading room. Two each with deep, shallow for the bottom, and then to their own in the subtle presentation of the opposite color. Such as Zhuwo study continuation of public space of the semicircular vocabulary, guide the visitors into the leather sofa, dark bookcase, stone mosaic structure of the heroic spirit field, but the use of light color carpet weaving, significantly increasing the space dynamic law; the second master's study entered into semi outdoor open design, with white floor bedding, but still use black bookcase with iron gray couch, eye-catching trumpet shaped lamps, concrete and subtle reflects the owner preferences, show both inside and outside the echo of the attitude to life.

For homeowners in the public-private boundary value, Xuanwu design will also be this concept into account, regional public gates using white, give close, sense of pure; into the private area in the black segment, with isolation, highlighting the "official" meaning. Into the secondary room, chess and card room and dining room divided according to left and right, the two are white as the keynote, checkerboard floor, placing classic crystal lamp, collocation Baroque pattern seats, dove gray pillows, far view close look, the deep rhyme.

In order to make the subjective straight from the comfortable, host and guest bedroom with consistent gentle color, again in different directions lines outline the space expression, such as concise Zhuwo elongated filament, and golden mattress, light blue carpet Xiangyingchengqu, reduce excessive stack a diffuse sense; the rest of the bedroom is to green water, blue sky is an ornament, in white, light gray tone, curtains, bedding and carpet slightly present, complicated with peony pattern bed backboards, compose a homeowner leisurely and elegant taste of life.

面朝大海 春暖花开
Facing the Sea in Spring

项目名称：深圳半山海景别墅设计
设计单位：朗昇空间设计
项目面积：1000 ㎡
项目地址：广东省深圳市
主要材料：大理石、地砖、涂料等

Project Name: Shenzhen Mid Ocean View Villa Design
Design Company: LONSON Design
Project Area: 1000 ㎡
Project Location: Shenzhen City, Guangdong Province
Main Materials: Marble, Floor Tiles, Paint, etc.

漂亮时光

我有一所房子
面朝大海，春暖花开
……

用知名诗人海子的《面朝大海 春暖花开》诗中这一句来形容深圳半山海景别墅设计再贴切不过了。我公司花费近一年时间设计的半山海景别墅于近期完全交付业主使用。该别墅设计共有5层，面积约为1000㎡，建筑主体依山而建，靠山面海，森林环抱，风景秀丽。室内则空间方正、南北通透、光线充足，可谓是一所真正意义上的面朝大海、春暖花开的房子。

因别墅是旧建筑，需要重新改造，朗昇空间设计团队为保持建筑与室内的统一与协调，将外观设计、园林设计、室内设计结合在一起来重新考虑。首先，建筑外观设计改动较大。用大理石雕刻的圆形拱门与几根巨大敦实、高高擎起屋脊的欧式石柱，显得十分稳重而有力。满铺的红褐色墙砖使建筑外形别具一格，欧式线条勾勒出建筑层次丰富的轮廓，使整个建筑外观看起来高贵、雅致，富有古典韵味。园林设计清新简洁，除了观赏鱼池及少量精心种植的小景观之外，预留了较多供家庭成员活动的空间。

客厅区域包含有主客厅、小酒吧、小会客厅等功能，由一、二楼两层贯通，楼层高、进深大，所以客厅设计尤为宽敞大气，层次丰富。客厅背景处两根巨大的石柱，似乎是建筑元素的延伸。一楼与二楼空间之间，通过几个圆拱造形相互连接，使上下空间互动起来。顶棚造型精心雕刻，悬挂的水晶灯，光彩夺目，华丽高贵。墙面上浅黄色涂料与白色面板相间，色彩丰富温暖。精美的欧式家具、地毯、窗帘及配饰物等，均饱含着业主的个性与喜好。这些饰物大多繁花似锦，似从户外盛开到室内，然后在每个空间、每个角落又以不同的形式自然绽放。

位于客厅一侧的餐厅设计，则是用建筑内一个看似封闭的圆形区域建造而成，其四周通透，光线充足，圆形的顶棚配以圆形的餐桌，顶棚上同样用雕刻的花纹构成，造型精致。其间垂吊的小型水晶灯，晶莹闪烁，搭配的黄色餐柜及灯饰，使餐厅中点缀着一丝异域的情调。

主卧室设计宽敞，景观视线尤佳，除摆放的双床榻之外，还摆放了一组休闲沙发茶几。主卧室设计风格与客厅区域设计风格几乎相同，从顶棚到床具、沙发等，亦是一番四季如春、繁花盛开的迷人景象，使主卧室充满着温暖与舒雅的气息。与主卧室配套的书房设计，在欧式风格的空间内，使用国人一贯喜欢的中式红木家具（书柜、书桌、床榻）等，散发出书香气息。

儿童房设计则显得清新淡雅，偶有几束淡淡的红色窗帘布艺，使室内充满着朝气与活力。由于家庭生活需要，本套别墅设计有两个家庭厅，分别设计成两种风格，两个家庭厅之间通过一扇长形花边窗户连接起来。一种为与别墅整体风格保持协调的欧式风格，这种风格似适宜于家庭中年龄较长的成员，具有私密感。另一种为东南亚风格，这种风格与其他休闲空间（如棋牌室设计、酒窖设计）等联系在一起，具有浓厚的热带度假气息，色质古朴厚重，有种令人回归自然的轻松休闲感，满足了家庭成员们娱乐及对外交际的需要。

豪宅软装十风格·优雅巴洛克

I have a house which is facing the sea
when spring is coming and temperature become warm
flowers will be blooming
……

The Half Mountain Sea View Villa Design project can be described by famous poet Haizi's poem Facing the sea with spring blossoms. Our company uses almost one year to design the project, now it is given to the owner. The villa has five floors, it locates 1000 ㎡. The main body of the building is near the mountain, the front is facing the sea. Lush forest and landscape are beautiful. Inner space is square shape will full air and light. It is really a house facing the sea in the blooming spring.

The house is old building which need to be re-transformed. Lang Sheng Space Design makes a combination of facade design, landscape design, inner design together to maintain the coordination of architecture and inner space. First of all, the facade has been changed a lot. Marble round door and some big and high European style column seems stable and strong. The red color brick makes the surface of the building unique. European lines stretch out rich outline of the building. The whole architecture is noble and elegant, full of classic sense. Garden design is fresh and simple, except for fish pool and some plants, it leaves more space for family members.

Living room space contains main guest room, bar, meeting room and other functions. The space has two floors. The storey is high, depth is wide. The living room design is wide and rich. The living room has two huge columns, it is the extension of architectural elements. Between the first floor and the second floor, there are some arch shape lines connecting the upper space and low space. The ceiling pattern has been carved by heart. The hanging crystal lamp is bright and noble. The wall has warm light yellow and white panel. European furniture carpet, curtain and ornaments contain owners' individuality and interests. Most of the decoration are like beautiful flowers from outer to inner, blooming in different space and different corner.

The dinning room design on the side of living room is built by a closed round area in the building. The surrounding is transparent. The light is sufficient. The round ceiling matches round table. The ceiling is full of carved pattern. The room is delicate, the small chandelier on the ceiling is bright. The yellow cabinet and light add some exotic tone to the dinning room.

The main bedroom is wide and has good view sight. Beside the double bed, there is a group of casual sofa table. The design style of the main bedroom and the living room are almost the same. From ceiling to bed and sofa, the flower pattern is very attracting, making the main bedroom full of warm and smooth sense. The study design uses European style. The Chinese rosewood furniture (bookcases, desks, couches, Chinese paintings) gives out book taste.

Children room design is fresh and elegant. The light red curtain fabrics makes the room full of vigor and vitality. Because of the need in family life, the captioned villa has been designed with two living rooms. They are designed in two styles respectively. The two families are connected with a long shape flower frame window. One style is European style which coordinates with the whole style of the villa, and this style is suitable for families' elder member. The style is full of private sense. Another style is Eastern and Southern style. This style is connected with other leisure space (Chess room design, Wine cellar design). The style is full of rich tropical resort ambience. The color is plain and thick, with a sense of getting close to nature, satisfying the need of family entertainment and external communications.

变奏华尔兹
Waltz Variations

项目名称：成都东大街样板房
设计公司：台北玄武设计
设计师：黄书恒、林胤汶
软装设计：胡春惠、张禾蒂、杨惠涵
项目面积：200 ㎡
主要材料：超白镜酸洗、白色钢琴烤漆、大花白大理石、
黑白根大理石、千层玉大理石
摄影师：王基守
撰文：程歆淳

Project Name: Chengdu East Street Show Flat
Design company: Taipei Sherwood Design
Designer: Huang Shuheng, Lin Yinwen
Soft Decoration Design: Hu Chunhui, Zhang Hedi, Yang Huihan
Project Area: 200 ㎡
Main Materials: Ultra White Mirror Pickling, White Piano Paint, Pickling Prabescato Marble, Black and White Marble, Thousand Jade Marble
Photographer: Wang Jishou
Author: Cheng Xinchun

豪宅软装十风格·优雅巴洛克

Hallway

Upon entering the hallway, your eyes would be attracted by the heavy and complicated parquet on the floor, dark marble sets off the pure white totem and the color contrast makes the space appear more three-dimensional. The hidden tiny cracks inside the stones are just like invisible clues, leading lines of sight extending upward all the way. Baroque style arch is accompanied with threedimensional paper cutting pasted on panes, with ardent rhythms of samba dance, and initiating the design prelude with dim light and shadow hidden indoors.

Living Room

Semi-circular classic pillar decorations inside the living room produce complete cylinders with the mirror reflections on both sides, with classic solemn feel and modern fashionable atmosphere. The heavy and complicated totem on the dining hall's wall is just like some "glass wallpaper," displaying innovative capacities of materials to the extremes. The ground interface of living room and dining hall makes use of the spectacular combinations of stones inlaying stainless steel. Winding metal patterns go all the way into the deep corridor, which not only divides the space, but also extends the space feel.

Bedroom

The master bedroom has sedateness and maturity as the design tone. The accessories of champagne, black and white pattern, bright silver and light gray colors can appropriately restrain the metal texture of the wall surface. Contrary to this, the children's room boldly applies blue color tone, echoing powerful and unconstrained imaginations of children.

豪宅软装十风格·优雅巴洛克

玄关

踏入玄关,目光就被地面上的繁复拼花所吸引,黝黑的大理石衬托着纯白的图案,色彩的对比使得空间更加立体。石材中隐带的微小裂纹,仿若暗藏的线索,引导视线一路延展而上。巴洛克风格的圆拱搭配着立体的窗花,有着桑巴舞的热情节奏,与潜藏于室内的隐约光影,共同开启设计的序幕。

客厅

客厅的半圆形古典柱饰,借由两旁镜面的反射,合成了一根完整的圆柱,既有古典的庄重感,也有现代的时尚气息。餐厅墙面繁复华丽的图案,仿若一张"玻璃化的壁纸",将材料的创意机能发挥到极致。客厅与餐厅的地面交界处,使用了石材镶嵌不锈钢的特殊组合,曲折的金属纹样直入深邃的走廊,既区分了空间,也延伸了空间。

卧室

主卧室以沉稳、成熟为设计的基调,香槟色、黑白纹、亮银、浅灰的摆设,能适度收敛壁面的金属质感;与此相反,儿童房大胆地采用了蓝色调,呼应着孩童天马行空的想象力。

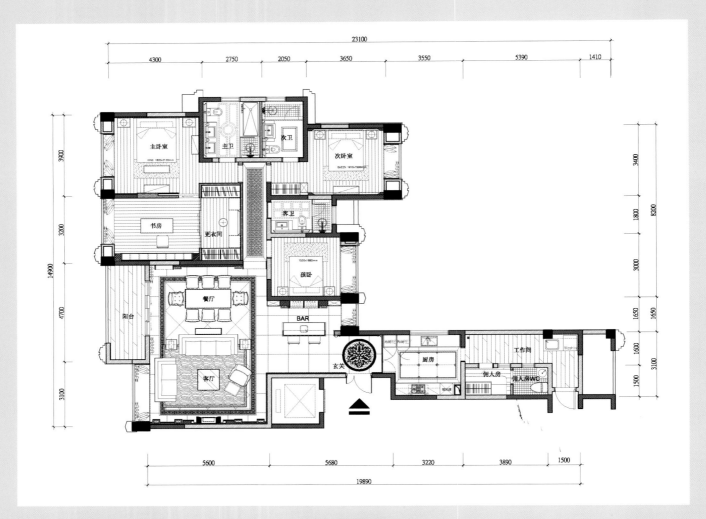

一层平面布置图

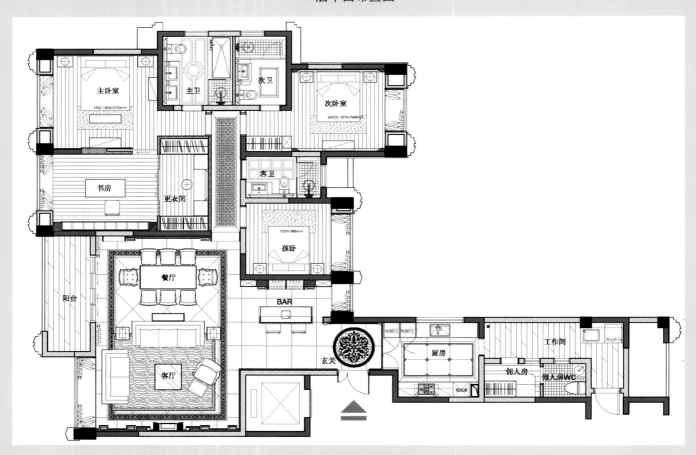

二层平面布置图

漂亮时光

浪漫法式
French Romantic

优雅孔雀蓝
Elegant Peacock Blue

项目名称：上海华贸·东滩花园项目 S1 户型
设计单位：上海壹陈建筑装饰设计工程有限公司
设计师：刘宏
项目地点：上海
项目面积：217 ㎡
主要材料：爵士白（石材）、清镜、白色烤漆、拼花木地板、金箔、艺术壁纸

Project Name: Shanghai China Trade - East Beach Garden Project S1 Type
Design Company: Shanghai ESSE SPACE Architectural Decoration Design Engineering Co., Ltd.
Designer: Liu Hong
Project Location: Shanghai
Design Area: 217 ㎡
Main Materials: White Jazz (Stone), Mirror, White Paint, Parquet Wood Floor, Gold Foil, Art Wallpaper

豪宅软装十风格·浪漫法式

豪宅软装十风格·浪漫法式

漂亮时光

本案以女性视角为主线，打造唯美、浪漫、精致、高贵的"服装设计师之家"。一所房子应具有气质和灵魂，这是长久以来壹陈空间所坚持的设计理念。这所面积为两百出头的小联排别墅，设计整体格局开放并且轻松，不刻意强调所谓的对称与空间气势，利用镜面、细部的雕花、石材马赛克拼花、水晶等材质的渲染，让空间显得层次丰富，反而呈现出大别墅的感觉。

穿插其中的孔雀蓝是整个设计的点睛之笔。在水晶与夜色的相映之下，仿若看见一位气质雍容典雅的少妇漫步在自己的工作室，若有所思，红酒摇曳，让人舍不得移开目光。

Take the female perspective as the main line, to create beautiful, romantic, elegant, noble "fashion designer's home", a house should have temperament and soul is a long time since the idea of a new space. The area of two hundred head of the small platoon villa, design the overall pattern of open and relaxed and enjoyable, not deliberately stressed the so-called symmetry and space style, and the use of the mirror, detail carved, stone mosaic, crystal material such as rendering, let a space is the level of rich, but showing a large villa feeling.

The peacock is interspersed among the design of the Punchline. Crystal and matched the night, imitate if saw a temperament elegant woman walking in their own studio thoughtfully, wine swaying, lets the human not averted their eyes.

漂亮时光

唯美蓝调
Aesthetic Blues Stone

作品名称：东营万芳园小区 B 户型别墅样板间
设计单位：上海壹陈建筑装饰设计工程有限公司
空间设计师：刘宏
软装设计师：吕庆楠
项目面积：251 ㎡
主要材质：爵士白大理石、黑美人大理石、拼花马赛克、拼花地板、实木复合地板、壁纸等
摄影：三像摄 张静

Project Name: Dongying Wanfang Park District Villa Show Flat Apartment Layout B
Design Company: Shanghai ESSE SPACE Architectural Decoration Design Engineering Co., Ltd.
Space Designer: Liu Hong
Soft Decoration Designer: Lv Qingnan
Project Area: 251 ㎡
Main Materials: Knight White Marble, Black Beauty Marble, Mosaic Pattern, Parquet Flooring, Solid Wood Flooring, Wallpaper, etc.
Photograph: Threeimages Zhang Jing

豪宅软装十风格·浪漫法式

豪宅软装十风格·浪漫法式

小区定位为别墅洋房居住区，为城市的高层次人士打造。本案项目的设计延续了售楼处的法式风格，摒弃了庄严恢宏的气势，在浪漫优雅上做足文章。

设计师在优雅浪漫的主基调下又营造出移步换景的空间美感，强调建筑绘画与雕塑以及室内环境等的平衡感，亦平衡贵气十足的感官冲击。绚丽的水晶散发出柔和的光芒，凝聚了偌大空间的情感，华丽却不落俗套。镜面的映射宛若水面的波纹相互交错，其柔性与动态，诠释了设计师对大宅精神的把控。

蓝色的家居软装为白色的空间里酝酿出浪漫矜贵的格调，亦是这唯美的蓝，将浪漫与高雅的法式风格渲染到极致。

儿童房清新的粉色，让整个空间充满着无限的生机和公主般的浪漫情怀。

Residential area for villa villa residential area, for the city to build a peak. The case for the sales offices of the French style, abandon the majestic momentum, make full text in romantic elegant.

Designer in the main tone of romantic and elegant and create a venue for space aesthetic feeling, emphasizes a comprehensive architectural painting and sculpture as well as indoor environment, but also balance the sensory impact of extreme extravagance, gorgeous matched crystal light, condensed the huge space of emotion, gorgeous but not vulgar. Mirror promenade, like water ripple crisscross and flexible and dynamic, show lead people to stop the luxury layout techniques, interpretation of the designer to house the spirit strength of the master.

Blue household soft outfit is white space brewing romantic boast commisserate expensive style is also this beautiful blue, the romantic and elegant French style rendering to the extreme.

Children's room fresh pink, so that the whole space is full of infinite vitality and Princess like romantic feelings.

漂亮时光

漂亮时光

豪宅软装十风格·浪漫法式

豪宅软装十风格·浪漫法式

桃花源
The Peach Garden

项目名称：上林西苑叠拼别墅样板房
设计公司：上海无相室内设计工程有限公司
空间设计师：王兵、王建
软装设计师：李倩
项目面积：250 ㎡
主要材料：雅士白大理石、古堡灰大理石、彩色烤漆、壁纸、马赛克
摄影：三像摄 张静

Project Name: Shanglin Xiyuan Diepin Villa Model Room
Design Company: Shanghai ANIMITTA Interior Design Engineering Co., Ltd.
Space Designers: Wang Bing, Wang Jian
Soft Decoration Designer: Li Qian
Project Area: 250 ㎡
Main Materials: White Marble, Castle Gray Marble, Color Paint, Wallpaper, Mosaic
Photography: Threeimages Zhang Jing

法式风格弥漫着复古、自然主义的格调，最突出的是贵族气十足。这种"贵"散发着人文和古典气息，舒适、优雅、安逸是它的内在气质。这种风格让人联想起庄园、钢琴、舞会、蓬蓬裙，精致化了的乡村风格，是人们一直都在追求的"桃花源"。

本案的设计展现出清新与优雅的氛围，注重人文气息的营造，儒雅中又充满了贵气。布局上突出轴线的对称与恢宏的气势，不求简单的协调，而是崇尚冲突之美。在设计上讲求心灵的自然回归感，给人一种浓郁的艺术气息。陈设注重区域组景构图，逐级递进，通过陈设加强整个空间的整体性。

本案的客厅是整个空间最具表现力的功能区域，细节处理上运用了经典的法式线条，制作工艺精细考究，特别是在家具及陈设品的表现效果上尤为明显。象牙白色的木饰面及丝织物的柔顺光滑，釉面瓷器瓶艺的晶莹剔透，浅色系的沙发与暗金质地的茶几组合配以浅灰色的真丝地毯、清新明快的抽象油画、典雅的水晶灯具展现出了一幅低调奢华、贵族气息十足的精彩画卷，浪漫清新之感扑面而来。

卧室中，不论是床头台灯图案中娇艳的花朵，抑或窗前一把微微晃动的摇椅，在任何一个角落，都能体会到主人悠然自得的生活和阳光般明媚的心情。

French style filled with retro, Mensao, naturalistic style, the most prominent is the noble gas. This "expensive" is a human and classical atmosphere, comfortable, elegant, comfortable is its inherent temperament. This style is reminiscent of the manor, piano, dance, Tutu, delicate rustic style, people have been in pursuit of "Utopia".

The design of the case to show the freshness and elegance, create a humanistic atmosphere, full of elegance and noble. Layout outstanding axis of symmetry, magnificent, luxurious and comfortable living space, not for simple coordination, but advocating the beauty of the conflict. In the design of the nature of the mind of the natural regression, giving a strong sense of art. Display focus on the regional group of landscape composition, progressive progressive, through the layout to strengthen the overall space.

In the case of the living room is the space the most expressive function area, the details of the deal using classical French lines, fine craftsmanship is exquisite, especially on the performance effects of furniture and furnishings. Ivory white wood veneer and silk fabric soft smooth, crystal clear glazed porcelain bottles arts, light coloured sofa and uniques texture combined tea table with to light gray silk carpet, fresh and bright abstract oil paintings, elegant crystal lamps show a picture of a low-key luxury, noble atmosphere full of wonderful picture scroll, romantic and fresh feeling blow on the face and come.

Bedroom space whether bedside table lamp pattern in the delicate and charming flowers or window to a slightly swaying rocking chair, in a corner, can experience to master leisurely life and sunshine bright mood.

富春江岸边的法式情怀
French feelings on the Shore of the Fuchun River

项目名称：绿城·富阳玫瑰园蓝玫苑法式样板房
设计公司：广州赫尔贝纳室内设计有限公司
设计师：钟志军
项目面积：340 ㎡
主要材料：圣罗兰大理石、罗马灰、西班牙米黄石、壁纸、青铜
摄影：三像摄 张静

Project Name: Fuyang Greentown•Fuyang Rose Garden Lanmei Yuan French Show Flat
Design Company : Guangzhou Herabenna Interior Design Co.,ltd.
Designer : Zhong Zhijun
Project Area: 340 ㎡
Main Material: Laurent , Rome Gray , Spain Cream-colored Stone, Wallpaper, Bronze
Photography: Threeimages Zhang Jing

漂亮时光

豪宅软装十风格·浪漫法式

一脉灵气的富春江，成就了"天下佳山水，古今推富阳"的美名，也吸引了一代画圣黄公望在此结庐定居，富春江两岸的经年游走，更成就了《富春山居图》的博然大气。富阳玫瑰园项目就坐落于此，毗邻黄公望隐居地，享受黄公望森林公园的天然资源，身处其中也能感受到深刻的文化内涵。

富阳玫瑰园项目分南入户型和北入户型2个户型，其中南入户型为时尚的法式风格，此设计以十八世纪的法国贵族生活为背景，结合当地的历史文化特色和精英阶层的生活方式，营造出一种低调奢华的生活场景，创造出一种宁静优雅的生活美学观。

该别墅定位为5口之家，男主人沉稳内敛，有着自己独特的审美观和敏锐的时尚触觉，女主人时尚浪漫，对时尚有自己独到的见解。主要空间有客厅、餐厅、卧室、书房、地下休闲区域。客厅营造出优雅内敛的气质，以黄色、奶油色为主色调，浅蓝色的点缀使整个空间清新而灵动。水晶吊灯光影重重，家具的线条精致优雅，如油画般的细腻笔触，将优雅感层层凸显出来。在这米白的基调中，精心雕琢的壁炉、简化的罗马柱头、古玩瓷器、丝绸刺绣在同一空间对话，共同完成着优雅写意的诉求，隐喻高雅谦逊的气度。

每个单独的空间都承担着自己的功能和内涵，又相互连通、心手相连、和谐统一。长辈房庄重沉稳，主卧文雅浪漫，书房谦逊内敛，儿童房天真浪漫，处处都在诉说着谦逊的生活信条，演绎着高雅浪漫的情怀。

The aura of a pulse of Fuchun River, the achievements of the "world best landscape, ancient and modern push in Fuyang reputation, also attracted generation Hua Sheng Huang Gongwang here Jielu settled rich spring river on both sides of the Taiwan Straits after years of wandering, but the achievements of the dwelling in the Fuchun mountains natural atmosphere. Fuyang rose garden project is located in the, adjacent to the seclusion of Huang Gongwang, enjoy the natural resources of Huang Gongwang Forest Park, living in them can feel the profound cultural connotation.

豪宅软装十风格·浪漫法式

The aura of a pulse of Fuchun River, the achievements of the "world best landscape, ancient and modern push in Fuyang reputation, also attracted generation Hua Sheng Huang Gongwang here Jielu settled rich spring river on both sides of the Taiwan Straits after years of wandering, but the achievements of the dwelling in the Fuchun mountains natural atmosphere. Fuyang rose garden project is located in the, adjacent to the seclusion of Huang Gongwang, enjoy the natural resources of Huang Gongwang Forest Park, living in them can feel the profound cultural connotation. Fuyang rose garden project is divided into South and North into the 2 Huxing, which is the south into the style of French style, the design of the French aristocracy living in eighteenth Century as the background, combined with the local historical and cultural characteristics and the elite life style, create a low-key luxury life scenes, creating a quiet elegant life aesthetics.

The villa positioning for five home, male master calm and reserved, their unique aesthetic view and a keen fashion sense, the hostess romantic fashion, fashion has its own unique insights. The main space has a living room, dining room, bedroom, study, underground leisure area. Living room to create a restrained and elegant temperament, mainly yellow, cream colored color, pale blue dot the whole space is pure and fresh and vivid. Crystal pendant light and shadow, the lines of the furniture is elegant, such as oil painting delicate brush strokes, will highlight the elegant layer. In this m white tone, crafted of fireplace, simplified Roman column, antique porcelain, silk embroidery in space with a dialogue together to complete a elegant freehand demands and metaphor elegant humble tolerance.

Each individual space undertakes the function and meaning of their own, and communicated with each other, Xinshouxianglian, harmony and unity. Elders real prudent and solemn, Zhuwo elegant romantic, study modest and restrained, children room is innocent and romantic, everywhere in recounting the humble life credo, interpretation of the elegant romantic feelings.

普罗旺斯法式风情
French Style in Provence

项目名称：布利杰印象巴黎样板房
室内设计公司：杭州易和室内设计有限公司
空间设计师：麻景进
软装陈设公司：杭州极尚装饰设计工程有限公司
软装设计师：沈肖逸、许彦文
项目地点：浙江省宁波市
项目面积：245 ㎡
主要材料：法国金花大理石、贝芝石尚大理石、混油饰面、壁纸、绒面硬包等

Project Name: Paris Bulijie Impression Show Flat
Interior Design Company: Hangzhou Yihe Interiors Design Co., Ltd.
Space Designer: Ma Jingjin
Soft Furnishings Company: Hangzhou Jishang Decorative Design Engineering Co., Ltd.
Soft Decoration Designer: Shen Xiaoyi , Xu Yanwen
Project Location: Ningbo City , Zhejiang Province
Project Area: 245 m^2
Main Materials: France Golden Marble, Becis Jean Marble, Mixed Oil Finish, Wallpaper, Hard Pack Flannel etc.

豪宅软装十风格·浪漫法式

布利杰印象巴黎别墅样板房，户型面积 245 m²，由设计师麻景进设计。不同于其他类型的样板房，设计师形象地诠释了"普罗旺斯法式风情"这种异域生活方式，将自然的、休闲的、艺术的法式贵族气息渗透在空间的每个角落，将人们在别墅内的生活细致而又精致地呈现出来。

设计师喜爱和尊崇法式别墅带给业主的贵族式生活的乐趣。整个空间色调的搭配营造出普罗旺斯居家生活的特质。浅淡的灰绿色穿插其间，还有优雅宁静的海蓝色来点缀，还有儿童房的森林色和那些玫红的、米黄的、金色的……无处不在的普罗旺斯色彩装点，让整个样板房充满了新鲜、阳光的活力。在这里处处可以找到讲究精美的轴线、墙面上的立体装饰线条和风情万种的配饰。

不仅如此，设计师还在地下室设置了台球室、酒窖和雪茄吧，营造出低调而从容的生活气氛，满足业主贵族式的爱好和交际需要。奢华于心才是极致的奢华，享乐于思才是至高的享乐，空间中呈现出的清雅法式格调，撇开了浓墨重彩的宫廷味，带给人们一种亲密无间的相融氛围。

Bulijie impression Vila Paris Hotel model room, apartment layout area of 245 m², designed by architect Ma Jingjin. Different from other types of housing model, designer image interpretation of "Provence style" of this different way of life, natural, leisure, art of French aristocratic infiltrated in every corner, the people in the villa life meticulous and exquisite present.

Designers love and respect to the aristocratic French villa owners of the joy of life. The entire space tonal collocation to create Provence home life qualities. Pale grayish green interspersed the meantime, there are elegant and quiet blue color to adorn and children of the housing forest color and those rose red, beige, golden... In Provence, everywhere in the color, so that the whole room is filled with fresh vitality, sunshine. Here can be found everywhere, on the walls of the axis of exquisite exquisite three-dimensional decorative lines and elegant accessories.

Not only that, designers also the basement set billiards room, wine cellar and cigars., to create the atmosphere of a low-key and quiet life, meet the owners of the aristocratic tastes and communication. Luxury in the heart just arrives too is the ultimate luxury, hedonic contemplation is the supreme enjoyment, space show the elegant French style, apart from the splendid palace flavor, bring people a close no from the integration of the atmosphere.

豪宅软装十风格·浪漫法式

诠释现代宫廷艺术
Interpretation of Modern Court Art

项目名称：华侨城某宅
设计单位：BKD – DESIGN
设计师：高翔
项目地点：湖北省武汉市
项目面积：335 ㎡
主要材料：法国灰大理石、雅士白大理石、黑白根大理石、拼花马赛克、贝壳马赛克、深沟纹橡木地板、拼花地板、进口壁纸

**Project Name: Overseas Chinese Town House
Design Company: BKD-DESIGN
Designer: Gaven Gao
Project Location: Wuhan City , Hubei Province
Project Area: 335 ㎡
Main Material: France Grey Marble, Aston White Marble, Black and White Marble, Mosaic Pattern, Shell Mosaic, Deep Groove Lines Oak Flooring, Parquet Floors, Imported Wallpaper**

所谓的法式风格不是简单延续古典的设计方法和形式。设计师在构思本案的时候，希望尊重传统的同时打破传统，在保留法式经典的同时，加入黑白灰的现代元素，让古典的装饰形式与现代的颜色产生碰撞，从而产生极强的视觉冲击力，给人深刻的印象。

手工的锻打铜吊灯、局部鎏金雕刻的家具、刺绣缎面的家具、精致花艺等元素的注入提升了整个空间的高贵感，呈现出一个具有现代气息的法式风格空间。

The so-called French style is not a simple extension of classical design methods and forms. Designers in the concept of this case, hope respect tradition and break the traditional, while retaining the classic French at the same time, adding modern elements of black, white and gray, ornamental forms of classical and modern color to produce collision, resulting in a strong visual impact, give people a deep impression.

Hand forged brass chandeliers, local gilt carved furniture, embroidered Satin furniture, delicate floral and other elements of the injection to enhance the noble sense of the whole space, showing a modern flavor of the French style space.

漂亮时光

豪宅软装十风格·浪漫法式

法式生活美学
French Life Aesthetics

项目名称：项目名称：绿城·台州宁江明月法式样板房
设计公司：广州赫尔贝纳室内设计有限公司
设计师：钟志军
项目面积：787 ㎡
主要材料：圣罗兰大理石、罗马灰大理石、西班牙米黄石、壁纸、青铜
摄影：三像摄　张静

Project Name : Greentown - Taizhou Ningjiang moon French Show Flat
Design Company: Guangzhou Herabenna Interior Design Co.,ltd.
Designer: Zhong Zhijun
Project Area: 787 ㎡
Main Materials: Classic Beige, Fantastic Color of Jade, Blue Diamond, Jay Liyakin, Moon Jade, Yellow Jade, Wallpaper, Mosaic
Photograph: Threeimages Zhang Jing

本项目位于浙江省台州市黄岩区的永宁江岸边，坐拥永宁江景，诠释出居住品质的新标杆。本案的设计以法式宫廷风格为主，融合了Chinoiserie的法式东方韵味，这种浓郁的宫廷色彩与贵族氛围，仿佛将我们带回那个辉煌的黄金时代。

推开厚重的入户大门，金色的富有雕塑感的装饰台、优雅的铜质吊灯，将你引入一个法式艺术生活的殿堂。

别墅共分为3层，首层主要是开放空间，客厅和家庭厅中间隔着花园，布局手法借鉴中式园林的设计法则。客厅以米白色为主色调，精致的雕花、细腻的描金，各色古玩摆件，在水晶灯的光晕下，折射出宫廷式的奢华与优雅感。中式元素的陶瓷凳配上明黄色的梅花装饰画，将传统中式元素与西方文化相融合，混搭出一种国际范儿。

高雅奢华的餐厅中，无论是一把刀、还是一个酒杯都经过设计师的精挑细选。阳光房中飘逸的窗帘像极了女王优雅的裙摆，而整个家庭室就像是女王的微笑，有着至高无上的尊贵面庞，却又不失暖人心怀的亲和力。在这秋天的午后，片片阳光洒进温馨的家庭室，倚在柔软的法式布艺沙发上，尽享一杯咖啡的香浓时光，浪漫就在这一点一滴中弥漫。

开敞的书房是体现主人品味的最佳选择，收藏的艺术品、书籍都在此彰显主人的高雅情趣。一层除了以上开放空间，还有两个卧室——长辈房和客房。长辈房温馨舒适，根据长辈的年龄特征，颜色选择的是相对沉稳的巧克力色和深蓝色，呈现的是一个温暖知性的空间氛围。客房的整体感觉简洁、舒适，局部点缀的蓝色，增强了空间的活跃度，也同其他空间形成呼应。

沿着楼梯拾阶而下，来到地下一层休闲娱乐区，高耸的穹顶悬挂着水晶灯，光影交相辉映，仿佛置身于十八世纪的宫廷宴会中。

楼上是相对私密的卧室空间。主卧运用轻柔的亮白色、奶油色，舒缓优雅的气质由内而外散发出来。精致的雕花大床，大块面的欧式花纹壁纸，再加上东方元素的融入，营造出一种让人迷恋的欧式格调。

The project is located in Huangyan District, Taizhou City, Zhejiang Province Yongning River shore, sitting on the Yongning River King, the interpretation of the new benchmark for the quality of living. The design of this case is the main style of the court style, the fusion of the Chinoiserie French oriental charm, this rich court color and aristocratic atmosphere, as if we bring back to the brilliant golden age.

豪宅软装十风格·浪漫法式

漂亮时光

豪宅软装十风格·浪漫法式

豪宅软装十风格·浪漫法式

漂亮时光

豪宅软装十风格·浪漫法式

漂亮时光

The villa is divided into 3 layers, the first layer is open space, the living room and family room in the middle of the garden, the layout of the Chinese landscape design. Living room with white rice based colors, exquisite carved, exquisite gilt, all kinds of antique ornaments, in the halo of crystal lamp, reflects the elegant. Chinese elements of ceramic stool with bright yellow plum decorative painting, traditional Chinese elements and Western culture fusion to mix and match an international range of children.

Open heavy doors into people's homes, gold rich sculptural decoration, elegant brass chandelier, leads you to a life of French art palace.

Elegant and luxurious restaurant, whether it is a knife, or a glass of the designer's fine selection. Sun room elegant curtains as the queen elegant skirt, and the family room like is smiling queen, supreme and dignified face, yet warm hearts and affinity. In this afternoon, patches of sunlight sprinkled into the warmth of the family room, leaning on the soft French fabric sofa, enjoy the fragrant time for a cup of coffee, romance is in the little bit of diffuse.

Open study is the embodiment of the best choice for the taste of the host or hostess, collection of artwork, books are here highlight the owner's elegant taste. In addition to the above open space, there are two bedrooms, the elders room and room. presents is a warm understanding space atmosphere. The overall feeling of the room is simple , the blue color of the local decoration, enhanced the space of activity, but also with other space form echo.

Along the staircase, to bear a layer of leisure and entertainment district, the soaring dome hanging crystal lights, lights add radiance and beauty to each other, as if exposure in the eighteenth century banquet in the palace.

Upstairs is a relatively private bedroom space. The master bedroom use gentle bright white, cream colored, soothing and elegant temperament from the inside out. combined with oriental elements into, creating a a people obsessed with the European style.

豪宅软装十风格·浪漫法式

漂亮时光

豪宅软装十风格·浪漫法式

豪宅软装十风格·浪漫法式

港铁天颂 大匠所筑
MTR Song Great Carpenter Built

项目名称：港铁荟港邸白地样板房 B3-03 户型
设计单位：SCD（香港）郑树芬设计师事务所
项目地点：广东省深圳市
项目面积：139 ㎡
主案设计师：郑树芬
参与设计师：杜恒、陈洁纯

**Project Name: MTR Hong Kong Hui Di White House B3-03 Apartment Layout
Design Company: SCD (HongKong) Simon Chong Design Consultants Limited
Project Location: Shenzhen City , Guangdong Province
Project Area: 139 ㎡
Main Designer: Simon Chong
Participated Designers: Amy Du , Chen Jiechun**

法国南部典型的中世纪小村——Gordes，被喻为"天空之城"，从宗教战争起一直保持着平和、简单的生活方式，这种生活态度一直被延续至今。Gordes的每个角落都流传着原始的味道，散发出古朴、厚实的质感。设计师将这种宁静美好融入设计中，造就朴实无华的生活空间。

本案最具特色之处，就是精致的设计细节与天然材料的和谐融合，创造出一个舒适自然的居住环境。客厅以木褐色为主色调，线织纹理的磨砂壁纸覆裹着客厅的墙壁，青木色厚麻布质感窗帘、人字形实木复合地板，勾勒出天然质朴的空间气质。富有光泽的牛皮棕色沙发、米色粗麻布面沙发与橄榄绿布面提花单人沙发相组合，造就了质感丰富的混搭效果。明媚多彩的花簇将法国南部旖旎的乡村气息引入空间，点亮了空间的色调，由内而外地散发出法国乡村风情的气质。

餐厅摆放着不经雕琢打磨的木制餐桌，造型简单、线条直接，原生态的纹理烙印着岁月流转的痕迹，木质餐椅则增加了绸面提花的细节，同种材质的原生态木柜在各个功能空间也能觅得踪影。由厨房推拉门演变而成的红木酒柜，及高脚杯形水晶灯让餐厅多了一分雅致的情调。

主卧的每个细节都值得细细品味。流行于路易时期的经典法式印花棉布出现在床品、床头软包、床尾凳上，地毯的图案及蓝青色调与此相呼应。墙面上的大幅挂画恬淡轻柔，符合卧室温馨浪漫的气质，摆饰、细节上的贯通让整个主卧饱满而和谐。

淡黄色的条纹壁纸给儿童房带来活力与朝气，运动元素的融入塑造了小男孩活泼好动的鲜明个性。色调清淡的花卉床品为次卧带来清新之感，加入柠檬黄的点缀，让空间更明朗。浴室以白色、淡黄色色调为主，质朴大方，马赛克的运用则增强了法式乡村的韵味。

这里真切地造就了法式乡村所强调的宁静与悠闲，轻松舒适的环境氛围引人回归朴素雅致的本真生活。

French Southern typical medieval village - gordes has been hailed as "the city of the sky", from the wars of religion has maintained a peaceful, simple way of life, away from the hustle and bustle of the flashy life attitude has been continued ever since. Each corner of the Gordes has been circulating the original flavor, emitting a simple, thick texture. The French designers will be strong trait of serene and beautiful into the design, to create the living space of chastity.

The most characteristic of the case is that the design details and the harmony of the natural materials, and thus bring out a comfortable living environment. Living room with wood brown tones, thread woven texture matte wallpaper cover wrapped in the living room walls; Aoki color thick linen texture curtains, herringbone wood composite floor, draw the outline of natural simple and elegant temperament. Full of shiny brown leather sofa, beige coarse linen cloth sofa and olive green cloth jacquard sofa combination, created a rich texture of mix and match effect. Bright and colorful flowers will into space in the south of France, and charming rustic, lit up the space of color, from the inside while the field distribution of the temperament of the French country style.

Restaurant decorated with not the carved polished wooden table, modeling simple, line directly, the original ecological texture bears the traces of years of circulation, wooden chair increased jacquard silk surface details, the same material of original ecological wood cabinets in various functional spaces can also find the trace. From the kitchen sliding door evolved mahogany red wine goblet, and crystal lamp to the restaurant a elegant atmosphere.

Master bedroom every detail deserve to savor, popular appeared on the bed, the head of a bed of soft bags, bed tail stool in the period of Louis the classic French calico, carpets and Lanqing tone and the echoes; on the walls of a hanging painting tranquil gentle, in line with the bedroom warm and romantic temperament, ornaments, details through let the master bedroom is full and harmonious.

Pale yellow stripe wallpaper to children room bring vitality, motion of the elements into shape the individuality of the lively little boy. The flower bed with light color is light and fresh feeling, add lemon yellow dot, let the space more bright. The bathroom with white, light yellow color, simple and easy, the use of the mosaic of the French countryside.

Here to create a truly peaceful and leisurely, relaxed and comfortable environment for people to return to the real life.

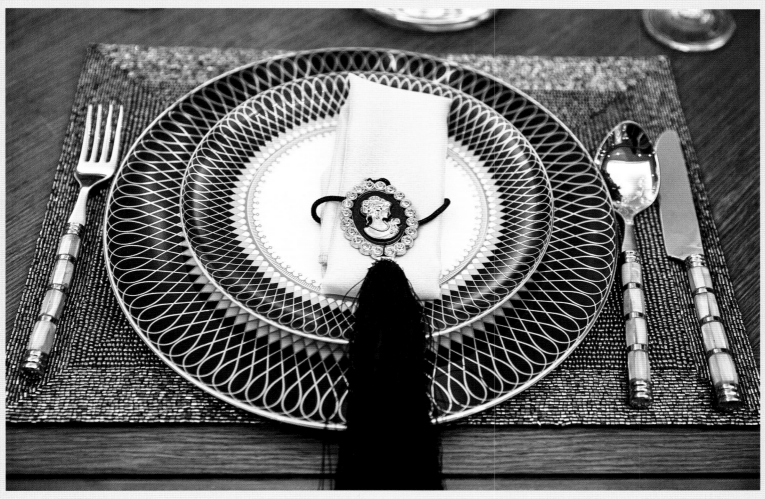

豪宅软装十风格·浪漫法式

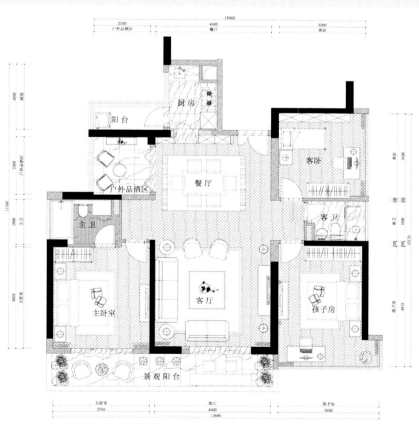

平面布置图

岁月静好　陌上花开
Years of Quiet Good, Mo Flowers

项目名称：荣禾·曲池东岸二期C户型
设计单位：SCD（香港）郑树芬设计师事务所
主案设计师：郑树芬
参与设计师：杜恒、陈洁纯
项目地址：陕西省西安市
项目面积：242 ㎡
摄影师：叶景星

Project Name: Ronghe - Quchi East Bank two Phase C Apartment layout
Design Company: SCD (HongKong) Simon Chong Design Consultants Limited
Main Designer: Simon Chong
Participated Designers: Amy Du , Chen Jiechun
Project Location : Xi'an City , Shanxi Province
Project Area: 242 ㎡
Photographer: Ye Jingxing

豪宅软装十风格·浪漫法式

与沙发同宽的棉麻刺绣布画是客厅的主角,画面丰富细腻,色调安静温暖,与灰蓝色的地毯相呼应。天然质朴的亚麻色调成为主色调,各个空间的色调既相成一体,又通过一些造型别致、富有质感的原木家具丰富了色彩的层次感。

无论是客厅还是家庭厅,在搭配方面非常讲究,细节之处散发出浪漫的气息。色彩柔和的西洋风情图案地毯、造型优美的法兰西瓷器、棉麻刺绣布艺、花卉床品等,都散发着淡淡的优雅,清浅的浪漫气息氤氲在空间每个角落。

客厅中,手感舒适的刺绣布艺与同色系的大理石地板相搭配,造就了温馨愉悦的就餐氛围。

主卧亦散发着浪漫的气息,浅色调的玫瑰花壁纸,花卉床品,白色流线型木柜和柔美的陶瓷摆件,如法国少女般温婉。小汽车、棒球、探险等都是小男孩的最爱,天蓝色的儿童房将这些都收容进去,建造起一个快乐的生活空间。温和的暖黄色调让次卧充满舒适感,亚黄色的法式花鸟图床品,墙壁上的花卉画都散发着淡淡的优雅。

整个空间延续了法式田园风格的天然一成、清浅浪漫的格调。没有过多的华丽装饰,反倒用自然简约的线条、色彩来营造出田园般舒适轻松的氛围。法式软装元素融进简朴的空间,谱唱出典雅浪漫的空间曲调。

And sofa with wide cotton embroidery fabric painting become the living room of the protagonist, the picture is exquisite and rich, tone quiet and warm, and grey blue carpet echoes. Natural and plain linen color to become the main colors, various space tone is complementary to one and by several modeling chic, rich texture of the wood furniture rich color level.

Whether it is the living room or family room, in collocation is very particular about the details of the distribution of a romantic atmosphere. Soft colors of western style carpet pattern, elegant French porcelain, cotton embroidery fabric, flower bed goods exudes a touch of elegance and shallow romantic breath dense in every corner.

Beige light texture details, feel comfortable embroidery fabric chairs, matched with the same color veins marble floor, creating a warm and pleasant dining atmosphere.

Zhuwo also exudes a romantic atmosphere, light colored roses wallpaper, flower bed, the sleek white wood cabinets and soft ceramic ornaments, such as French girl like gentle. Cars, baseball, adventure, and so are the little boy's dream, the sky blue children's room will be all these little dreams, for him to build a happy life space. Mild warm yellow tone let Ciwo with comfort, yellow sub French flower bed goods, wall paintings of flowers exudes touch of elegance.

The entire space continuation of the French garden style into a natural, clear romantic style. Not too many ornate decoration, but instead of natural simplicity of the line, color to create a pastoral and comfortable atmosphere. French soft mounted elements into the space spectrum of Jane Pu, singing tune elegant and romantic space.

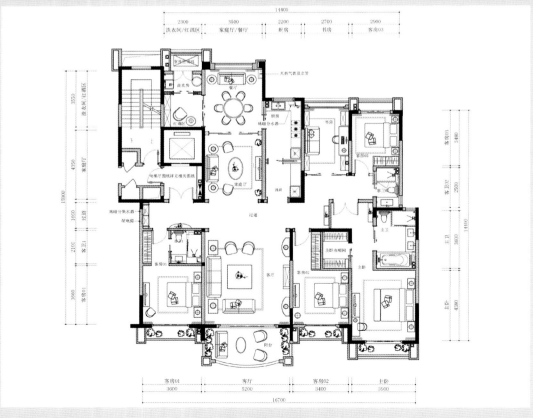

平面布置图

豪宅软装十风格·浪漫法式

西班牙新古典
Spain's New Classical Style

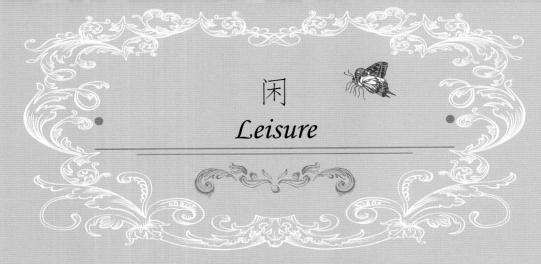

闲
Leisure

项目名称：新疆·中航翡翠城联排示范单位
设计公司：PINKI 品伊国际创意 & PINKI DESIGN 美国 IARI 刘卫军设计事务所
设计师：刘卫军
参与设计：梁义、卢浩、方永杰
项目面积：304 ㎡
项目地点：乌鲁木齐新市区国家级生态旅游区
主要材料：仿古橡木饰面板、仿古砖、机理木地板、石材、壁纸
开发商：中航地产

Project Name: Xinjiang . China Aviation Jade City Platoon Demonstration Unit
Design Company: PINKI Iran International Creative & PINKI DESIGN American IARI Liu Weijun Design Firm
Designer: Liu Weijun
Participated Designers: Liang Yi, Lu Hao, Fang Yongjie
Project Area: 304 ㎡
Project Location: Urumqi New Urban National Ecological Tourist Area
Main Materials: Antique Oak Decorative Panels, Antique Tiles, Mechanism of Wood Floors, Stone, Wallpaper
Developers: Aviation Real Estate

在本案中，整体空间舒适、典雅，泛着宁静的绿意，和着室内精炼的装饰细节，柔软而充满诗意的氛围，弥漫在整个空间中，引领着另一种独特的别墅生活。

在整个独特的空间氛围中，人们追忆生活的场景片段，沐浴着晨曦的光辉，自然中透着闲逸，听闻窗外鸟儿伴着和风飞舞，感叹生活竟如此美好。

Overall comfort and elegant green, glowing noble Shiba, and the refined interior details, soft and suffused with poetic, diffuse in the atmosphere, leading another unique villa life imagination.

In the unique space rendering, recalling the life scenes, bathed in the dawn light, nature reveals a leisurely and heard the birds outside the window with and danced in the wind. The same is true of life.

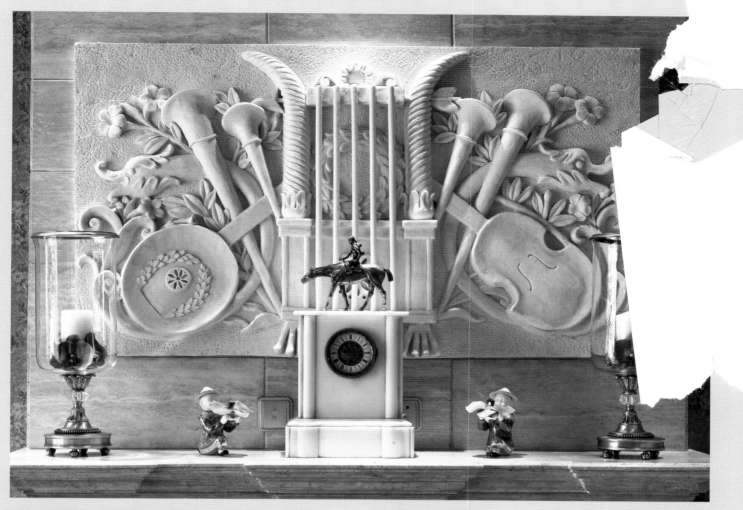

豪宅软装十风格·西班牙新古典

豪宅软装十风格·西班牙新古典

托斯卡纳艳阳
Tuscany Sunny

项目名称：广州南沙星河丹堤别墅
设计单位：深圳市孟伟室内设计有限公司
设计师：孟伟
项目地点：广东省南沙市
项目面积：367 ㎡
主要材料：大理石、布艺、手绘、布板，墙纸、玻璃
摄影师：江国增

Project Name: Guangzhou Nansha Galaxy Dante Villa
Design Company: Shenzhen Meng Wei Interior Design Co., Ltd.
Designer: Meng Wei
Project location: Nansha City, Guangdong Province
Project Area: 367 ㎡
Main Materials: Marble, Cloth, Hand-Painted, Cloth Board, Wallpaper, Glass
Photographer: Jiang Guozeng

本案设计风格是时下流行的托斯卡纳风格，托斯卡纳风格是乡村的、简朴的，但更是优雅的，它是室内设计与大自然的有机结合。人们会陶醉于其间的甜美与浪漫，也会欣赏它的纯朴与自然。

整个空间的主色调为暖黄色，仿佛托斯卡纳的艳阳照耀下的美丽小镇，散发出迷人的光辉。设计师在空间色调的设计上为上浅下深，重视室内空间的使用效能，强调室内布置按功能区分，主张废弃多余的、繁琐的附加装饰，追随流行时尚的装饰色彩及造型，打造独一无二的装饰空间。设计师在家具配饰上选用意大利风格的古典家具，且多带有花纹、雕饰，造形完美精致又兼具奢华的特质。

进入客厅，首先映入眼帘的是带有典型欧式风格特征的壁炉，岩石与灰泥的结合也是整个住宅的精髓之一。天花上的木结构体现出庄重感，铁艺的麦穗纹样灯具以及漂亮的彩砖瞬间将人们带入了意大利的古朴乡村中。门窗利用拱与柱相结合的方式体现出意大利建筑的形式美，又以浪漫的彩绘加以点缀，营造出高贵典雅、流光溢彩的氛围。

豪宅软装十风格·西班牙新古典

This case design style is the popular Tuscany style, Tuscany style is the village, simple, but more elegant, it is the organic combination of interior design and nature. People will be intoxicated with the sweet and romantic, but also appreciate its simplicity and nature.

The space of the main tone for the warm yellow, as if the Tuscan sun shine beautiful town, emitting a charming glory. Designers in the design of space colors for the shallow depth, attention to the use of indoor space, the indoor layout according to the functional distinction, advocate the use of redundant, cumbersome additional decoration, to follow the fashion of the decorative color and shape, to create a unique decorative space. Designers in the furniture accessories selected Italian style of classical furniture, and with patterns, carving, the shape is perfectly fine and combines the characteristics of luxury.

Entered the living room, the first thing that catches the eye is a typical European style fireplace. The essence of rock and mortar is a combination of the whole house. Smallpox wood structure reflects the dignified sense, iron grain pattern lamp and beautiful color brick instantly people into the quaint village of Italy. Doors and windows use arch and column combination way reflects the formal beauty in architecture in Italy, the romantic painting embellishment, to create a noble and elegant, Ambilight atmosphere.

豪宅软装十风格·西班牙新古典

爱在深秋的稻谷"金"
Love in the Autumn "Golden Rice"

项目名称：香港南益·紫兰湖国际高尔夫别墅示范单位
项目面积：720 ㎡
设计单位：香港方黄建筑师事务所
设计师：方峻
摄影师：江国增

Project Name: HongKong Nanyi · Purple Blue Lake International Golf Villa Demonstration Unit
Project Area: 720 ㎡
Design Company: HongKong Fong Architects & Associates
Designer: Noah Fong
Photographer: Jiang Guozeng

我们的灵感取自呈现丰收之感的稻谷的"金"色。这种色彩很容易令人联想到收获的季节，传递出丰富、光辉和温暖的气息。

在欧洲，这种颜色名为"Sunorange"，同样有丰收之意，从罗马时代开始就被当作多产的象征。因此，这种色彩无论在东方亦或在西方，都象征着富贵与幸福。

We were inspired by the "golden" color of the rice that was in the sense of the harvest. This color is very easy to think of the harvest season, pass out the rich, brilliant and warm atmosphere.

In Europe, the name is "Sunorange", as well as the meaning of the harvest, from the beginning of the Rome era as a symbol of fertility. Therefore, this color in the east or in the west, is a symbol of wealth and happiness.

漂亮时光

豪宅软装十风格·西班牙新古典

豪宅软装十风格·西班牙新古典

高贵与浪漫的异国风情
Noble and Romantic Exotic

项目名称：新昌悦澜山样板房
室内设计公司：杭州易和室内设计有限公司
空间设计师：李扬
软装设计公司：杭州极尚装饰设计工程有限公司
陈设设计师：李善洲
项目地点：浙江省绍兴市
项目面积：392 ㎡

Project Name: Xinchang Yue Lan Shan Show Flat
Interior Design Company: Hangzhou Ehe Interior Co., Ltd.
Space Designer: Li Yang
Soft Decoration Design Company: Hangzhou Artmost Decoration Design Engineering Co., Ltd.
Display Designer: Li Shanzhou
Project Location: Shaoxing City , Zhejiang Province
Project area: 392 ㎡

设计师用一贯擅长的设计语言巧妙还原高贵与浪漫的西班牙异国风情。西班牙家居带有强烈的地中海风格，热情洋溢、自由奔放、色彩绚丽。相对于地中海风格，更显得更神秘内敛、沉稳厚重。设计手法上，不需要讲究太多的技巧，而是保持简单的信念，捕捉光线、取材于自然，大胆而自由地运用色彩、造型。石材仿古处理，完全再现真正的地中海风情，古典铁艺类家具陈设和墙面贵族人物装饰画体现出庄园文化的传承。地下层大开间的书房，有收藏展示的功能，体现出主人的社会地位与品位。整个空间，散发出地中海生活的浪漫与自然情调，并且从骨子里透露出贵族般的高傲与奢侈。

豪宅软装十风格·西班牙新古典

豪宅软装十风格·西班牙新古典

Designers have always been good at the design language of the noble and romantic style of the romantic. The Spanish Home Furnishing with strong Mediterranean style, free spirited, colorful, ebullience. Relative to the Mediterranean style, more introverted, more mysterious thick and calm. Design approach, do not need to pay attention to too much of the skills, but to maintain a simple belief, to capture the light, based on the nature, bold and free to use color, shape. Stone archaizing treatment completely recreate real Mediterranean style, classical, wrought iron furniture and furnishings and wall aristocratic figure decorative painting reflects the inheritance of culture of manor. Underground layer of large bay study, exhibition, wine and cigar etc. to demonstrate their social status and taste. The entire space, exudes romantic and natural flavor of Mediterranean life, and from Pride and luxury with noble bones.

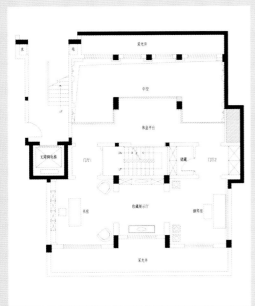

地下一层平面布置图

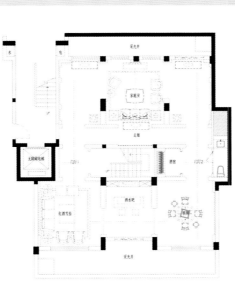

地下二层平面布置图

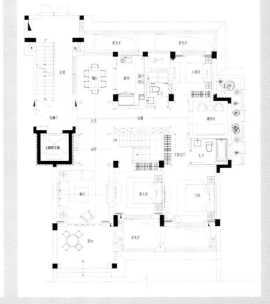

一层平面布置图

地中海度假
Mediterranean Resort Style

巴塞罗那的记忆
The Memory of Barcelona

项目名称：大禹·南湖首府185示范单位
设计公司：PINKI 品伊国际创意 & PINKI DESIGN 美国 IARI 刘卫军设计事务所
设计师：刘卫军
参与设计：梁义、袁朝贵
陈设公司：PINKI（品伊国际创意）&PINKI DECO 知本家陈设艺术机构
项目面积：185 ㎡
项目地点：吉林省长春市
主要材料：大理石、木饰面、壁纸、马赛克拼图

Project Name: Dayu, South Lake 185 Demonstration Units
Design Company: PINKI Iran International Creative & PINKI DESIGN American IARI Liu Weijun Design Firm
Designer: Liu Weijun
Participated Designer: Liang Yi , Yuan Chaogui
Display Company: PINKI (PINKI International Creative) & PINKI DECO Knowledge Owners Display Art Institutions
Project Area: 185 ㎡
Project Location: Changchun City , Jilin Province
Main Materials: Marble, Wood Veneer, Wallpaper, Mosaic Puzzles

每个人的心中都有一首蓝白色彩的诗歌。灿烂的阳光、金色的沙滩、蔚蓝色的海洋、碧蓝的天空、迎面而来的清风。带你领略一种简单无忧、轻松慵懒的生活方式，一种宠辱不惊，闲看庭前花开花落；去留无意，漫随天外云卷云舒的闲适意境。享受大自然赋予的美景，陶醉在巴塞罗那的浪漫情怀中。

Everyone's heart have a poem blue and white color. Brilliant sunshine, golden beaches, blue sea, blue sky, the oncoming wind. Take you enjoy a simple and worry-free, easy lazy way of life, be yourself, carefree look at the courthouse flowers bloom; Whether not, overflow with clouds scud across heaven leisurely and artistic conception. Enjoy the beauty of the nature, revel in the romantic feelings of Barcelona.

豪宅软装十风格·地中海度假

漂亮时光

豪宅软装十风格·地中海度假

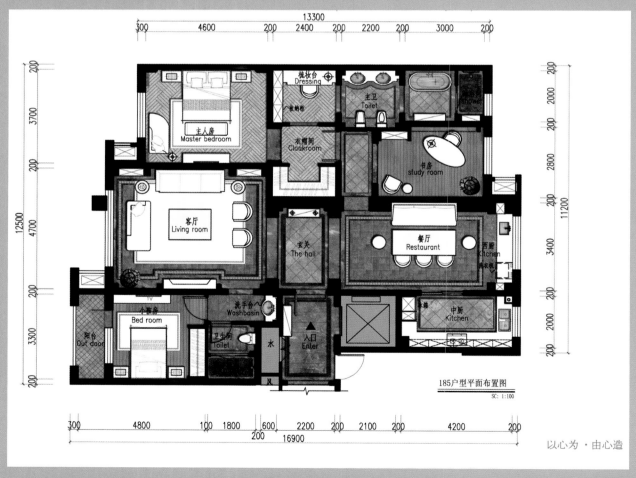

平面布置图

情迷地中海
Discovering Mediterranean

项目名称：紫悦府A户型别墅
设计公司：深圳市昊泽空间设计有限公司
设计师：韩松
项目地点：河南省洛阳市
项目面积：440 ㎡
主要材料：机理涂料、瓷砖、洗白木、大理石、石材马赛克

Project Name: Purple Yue Fu A type Villas
Design Company: Shenzhen HORIZON Space Design Co., LTD
Designer: Han Hong
Project Location: Luoyang City, Henan Province
Project Area: 440 ㎡
Main Materials: Mechanism Paint , Ceramic Tile, Wash White Wood, Marble, Stone Mosaic

为什么有这么多人迷恋地中海的感觉，因为地中海有着欧洲文明中最传奇的历史，特别是去了法国、意大利、西班牙，才真切地感受到了大部分东方人在视觉印象里感受不到的地中海气质：她很随意、散漫，因此很多时候是平静和浪漫的；她热情，奔放不羁，因此无拘无束，充满原始质朴的张力，她充满强大的包容力和融合性，因此能催生出最顶尖的艺术和生活……

这是我眼中的地中海，温和但充满传奇。

Why there are so many people obsessed with the feeling of the Mediterranean, because the Mediterranean has a history of European civilization, the most legendary, especially went to France, Italy, Spain, the only real feeling most of the children of the east in the visual sense of the Mediterranean temperament: she is very casual, lax, so most of the time is quiet and romantic; Her enthusiasm, bold and unrestrained unruly, so freely, full of the tension of primitive simplicity, she filled with strong capacity and fusion, and can produce the top art and life...

This is me in the eyes of the Mediterranean, mild but full of legend.

豪宅软装十风格·地中海度假

豪宅软装十风格·地中海度假

豪宅软装十风格·地中海度假

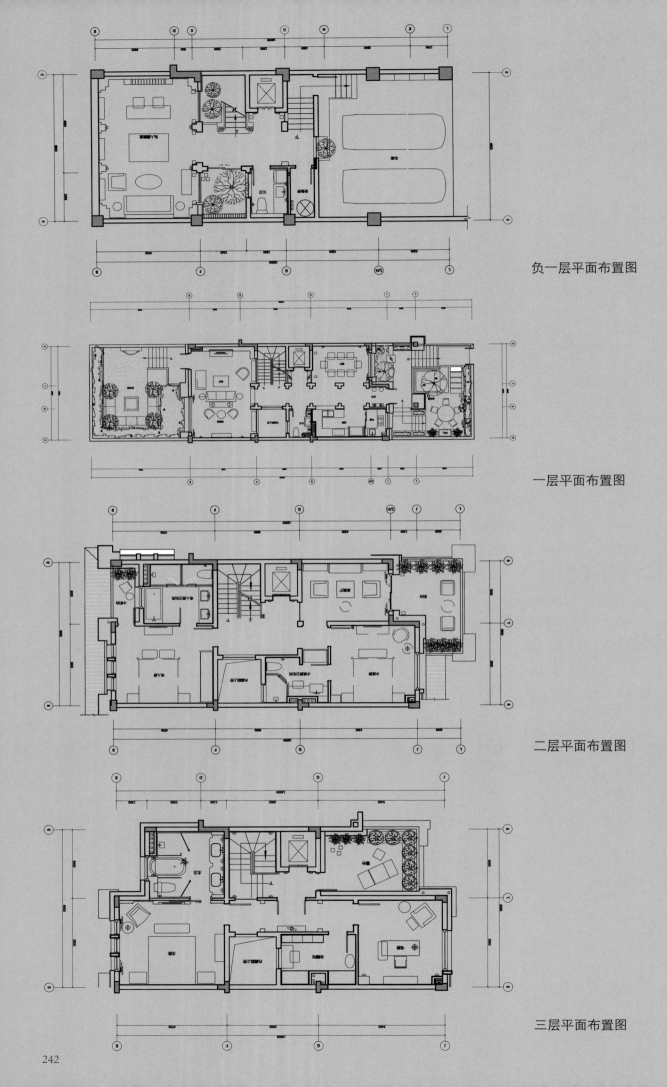

负一层平面布置图

一层平面布置图

二层平面布置图

三层平面布置图

豪宅软装十风格·地中海度假

豪宅软装十风格·地中海度假

豪宅软装十风格·地中海度假

蓝白迷情
Blue and White Rush

项目名称：建宇·雍山郡某宅
设计单位：重庆品辰装饰工程设计有限公司
设计师：庞一飞、殷正莉
项目地点：四川省重庆市
主要材料：雅士白大理石、机理涂料、壁纸、手工砖、铁艺、木地板

Project Name: A House of Jianyu · YongShan County
Design Company: Chongqing Pinchen Decoration Engineering Design Co., LTD
Designers: Pang Yifei, Yin ZhengLi
Project Location: Chongqing City, Sichuan Province
Main Materials: Scholars and White Marble, Mechanism Paint, Wallpaper, Manual Brick, Wrought Iron, Wood Floor

豪宅软装十风格·地中海度假

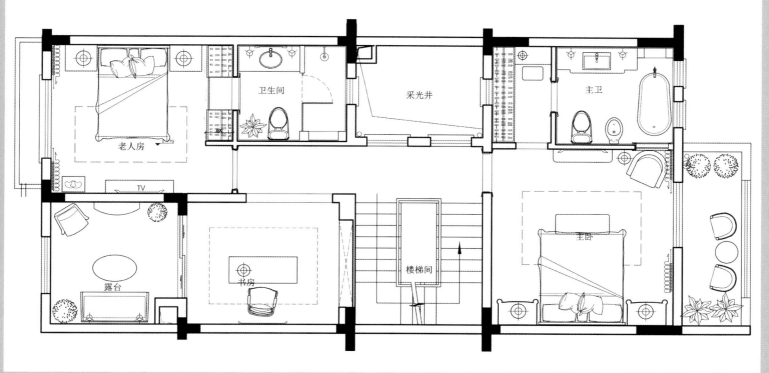

平面布置图

豪宅软装十风格·地中海度假

本案的设计以蓝白地中海风情为主调,温暖的光照、木纹肌理等元素使人在心理上,感受到开朗、活跃的氛围。设计师是物质和精神环境的创造者,不但应关心人的物质需要,更要了解人的心理需求。

本案的设计给人以如沐海上微风的浪漫感觉,使人感受到浪漫的气息。设计师恰如其分地组织空间结构,从单个空间的设计到群体空间的序列组织,由里到外,使得空间组织达到科学性、艺术性、理性与感性的完美结合。

客厅是进入居室的第一个主题空间,设计师运用镜中镜的形式在空间中创造层次感,中轴对称的细节布置、精心搭配的吊灯让人回忆起经典爱情电影的浪漫场景。美酒佳人,花前月下,地中海风情的陈设摆件烘托出罗曼蒂克的恋爱感觉。

卧室空间温馨惬意,用白色和蓝色来调节空间的视觉重量。光线结合色彩、质感、图形,把与空间功能匹配的体量美感和使用功能,艺术般地呈现出来,由此表达出浪漫的生活情怀,这就是本案例所阐述的空间气质。

The project design gives priority to tone with blue and white Mediterranean amorous feelings, warm light, wood texture, and other elements and gw YongShan county, interior space photograph echo, on the psychological effects, to achieve the effect of a bright and cheerful disposition, active. Stylist is the creator of the material and spiritual environment, not only should be concerned about the material needs of the people, even more to understand people's psychological needs.

Give a person with the case of design such as liquid, romantic feelings of the sea breeze, make people feel romantic breath. Designer appropriately organization space structure, from the design of a single space to group space sequence of organization, from the inside out, make the space organization to achieve the perfect combination of scientific, artistic, reason and sensibility.

The sitting room is to enter the first topic space of the bedroom, stylist is applied in the mirror mirror in the space to create the form of administrative levels sense, in the details of the axisymmetric decorate, elaborate collocation droplight reminiscent of the classic romantic love movie set. Wine wind, flowers, the display of the Mediterranean amorous feelings of furnishing articles of great romantic love feeling.

Warm and comfortable bedroom space, with white and blue to adjust the visual weight of space. Light color, texture, graphics, the function matching dimension and space aesthetic feeling and use function, the present like art, which expresses the romantic feelings of life, this is the case illustrated space temperament.

理想的生活 静静的靠近
The Ideal Life, Quiet Near

项目名称：大理某宅
设计单位：重庆品辰装饰工程设计有限公司
空间设计师：庞一飞、袁毅
软装设计师：张婧、夏婷婷
项目面积：180 ㎡
主要材料：做旧实木地板、硅藻泥、水曲柳木饰面、灰石材、麻布布艺

Project Name: A house of Dali
Design Company: Chongqing Pinchen Decoration Engineering Design Co., Ltd.
Space Designers : Pang Yifei, Yuan Yi
Soft Decoration Designers: Zhang Jing, Xia Tingting
Project Area: 180 ㎡
Main Materials: Do Old Real Wood Floor, Diatom Ooze, Ash Wood Veneer, Gray Stone, Linen Cloth

冬日暖阳，甜点搭配日光，坐在户外，看着飞鸟白云。光是这样呆呆地望着，心情就会很好。隐隐约约可以看到不远处的炊烟和昨日泛舟的洱海。这样的空间纵享大理的所有，能让人静下心来慢慢品味。

设计师将半地下的空间关系重新梳理，目的是让可以看见的柔和日光渗入室内，让人忘忧。

策划一个理想的下午，与悠闲一起散步。逛逛当地的菜市场，亲自为亲人或者朋友挑选食材，准备丰盛的一餐。可以发现生活中难以发现的世界，酝酿出许多鲜活的灵感，让创意能量不断累积。

定制的波斯地毯，羊皮手工灯，室内的暖色光线，让人想窝在室内。多少次到大理，新鲜感的期望值，已被它不断提升，感觉总要看到些许与众不同才能满足。

Winter warm sun, dessert with sunlight, sitting outside, watching the birds white clouds. Just so just looking at the mood will be very good. Can see a faint not smoke and white water rafting the erhai lake yesterday. All the space of longitudinal at Dali, no tourists ask, can let a person quietly taste.

Designers will be half basement space relationship, the goal is to make the soft sunlight into indoor can be seen, let a person worry.

Planning a perfect afternoon, and take a walk leisurely. Around the local market, as relatives or friends in person, choose the ingredients, preparing a meal. Can find life is hard to find in the world, making a lot of fresh ideas, make creative energy accumulating.

Customized Persian carpets, sheepskin lamp manually, indoor light color, let a person want to nest in the interior. How many times to Dali, expectations of novelty, has been rising, it feels total want to see a little different to meet.

豪宅软装十风格·地中海度假

漂亮时光

豪宅软装十风格·地中海度假

豪宅软装 十风格·地中海度假

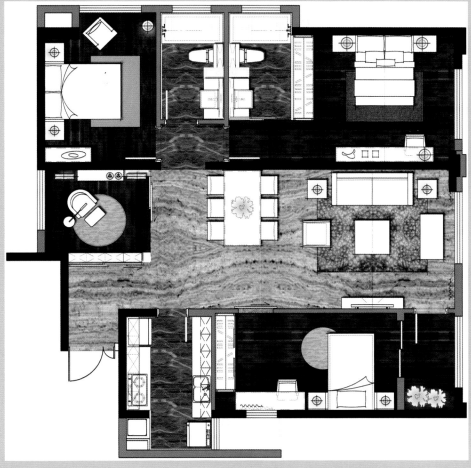

平面布置图

海的声音
The Sound of the Sea

项目名称：福州融汇桂湖生态温泉城样板房
设计公司：深圳市昊泽空间设计有限公司
设计师：韩松
项目地点：福建省福州市
项目面积：100 ㎡
主要材料：帕拉米黄大理石、洗白木饰面、胡桃木木地板、蓝色壁纸

Project Name: Fuzhou Integrated Gui Lake Ecological Model Housing of the Hot Springs
Design Company: Shenzhen HORIZON Space Design Co. Ltd.
Designer: Han Song
Project Location: Fuzhou City, Fujian Province
Project Area: 100 ㎡
Main Materials: Yellow Marble, Wash White Wood finish, Walnut Wood Floor, Blue Wallpaper

本案以地中海风格为主，完美演绎了整个空间的浪漫气息，大面积的蓝、白色调清澈无暇，诠释着人们对蓝天白云、碧海银沙的无尽渴望。

空间以海洋蔚蓝色的基底色调贯穿，蓝色调流动于整个空间中，充满着自然、浪漫的生活氛围。富有诗意般地勾勒出时代的气息，黑色铁艺、做旧的木色家具、亚麻质地的家居饰品，似乎都被这清澈的声音包裹着，融合相生。在任何一个角落，都能体会到主人悠然自得的生活和阳光般明媚的心情。

The case to the Mediterranean style, the perfect interpretation of the whole space, large area of blue and white spotless, interpretation of the people of the blue sky and white clouds, Bihaiyinsha endless desire.

Space Marine Wei blue basal tone throughout, blue adjusting dynamic in the whole space, full of natural, romantic breath of life. Poetic like to outline the era atmosphere, black iron and old wood color furniture, linen home accessories, appear to be the clear voice wrapped, the integration of carrier phase. In any corner, can feel carefree and content master life and sunny mood.

豪宅软装十风格·地中海度假

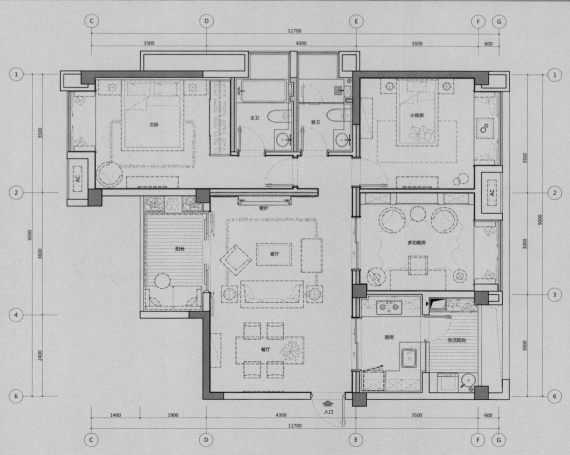

平面布置图

漂亮时光

豪宅软装十风格·地中海度假

复古雅皮
Retro Yuppie

复古的青春旋律
Retro Youth Melody

项目名称：中航城复式A2-5
设计单位：SCD（香港）郑树芬设计事务所
设计师：郑树芬
项目地点：贵阳
项目面积：206 ㎡

Project Name: China Aviation City A2-5
Design Company: SCD (HongKong) Simon Chong Design Firm
Designer:Simon Chong
Project Location: Guiyang
Project Area: 206 ㎡

精致的线条镌染着青春的轨迹，极具匠心的细致雕琢婉约地诉说着每一个指尖碰触过的唯美故事，或高雅尊贵，或雅致静怡。

这就是我们将要营造出的复古、前卫、混搭的惬意生活方式。

步入挑高大厅，一盏华丽的水晶吊灯，使空间充满了醉人的淡黄色灯光，富有时尚前沿的地毯，以驼色、黑色、白色交织出律动感，让复古混搭意境表现的活灵活现，展现着东西文化的融合，挑空的玻璃大窗与垂直的窗帘像一位优雅的女子披着金色的长发般美丽。沙发不仅以其古典的深灰色与实木框架结合，其舒适度也是设计师经过精心挑选过。

豪宅软装十风格·复古雅皮

豪宅软装十风格·复古雅皮

豪宅软装十风格·复古雅皮

豪宅软装十风格·复古雅皮

复古的青春旋律
Retro Youth Melody

项目名称：中航城复式A2-5
设计单位：SCD（香港）郑树芬设计事务所
设计师：郑树芬
项目地点：贵州省贵阳市
项目面积：206 ㎡

Project Name: China Aviation City Duplex Type A2-5
Design Company: SCD (HongKong) Simon Chong Design Consultants Limited
Designer: Simon Chong
Project Location: Guiyang City, Guizhou Province
Project Area: 206 ㎡

豪宅软装十风格·复古雅皮

豪宅软装十风格·复古雅皮

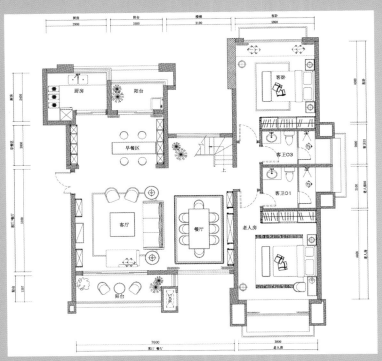

一层平面布置图

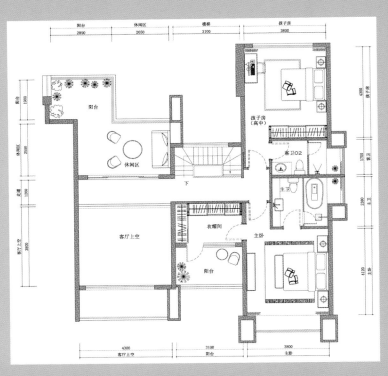

二层平面布置图

豪宅软装十风格·复古雅皮

豪宅软装十风格·复古雅皮

Crystal lamp restaurant is so noble and elegant, and its not only with living room chandelier caters to, the shape seems to have and the table is so beautiful and harmonious, classical European style dining chair seems to be the protagonist of the restaurant, the interpretation of a beautiful story. Restaurant side side of the closet display a variety of wine, red wine; spacious kitchen, especially to increase the breakfast table, of course, as the bar is elegant, appears elegant woman a beautiful encounter.

The second floor bright master bedroom, elegant bed with soft background wall, gray and brown bedside and unique bed goods left a space, subtle and complex printing carpets to the space added a heavy feeling. Beautiful linen pillow wall softening the entire space stiff feeling, let in the quiet yet luxurious bedroom. In the bathroom, the marble floors and walls, the room in a serious texture to be extended.

Next to the bedroom and children's room is a high school boy, the camel is a dark green, the modern and classical elements into one, carpet the full of vitality pattern and has the soft outfit embellishment of European elements, let the room to add a few single playful and lively.

餐厅的水晶灯是如此的高贵典雅，不仅与客厅的吊灯相迎合，其形状似乎也与餐桌显得如此的和谐，欧式古典的餐椅是整个餐厅的主角。餐厅一侧的壁柜展示着各式各样的洋酒、红酒；宽敞的厨房里，还特别增加了早餐台，当作吧台也不失其典雅。

二楼明亮的主卧，优雅的大床配以柔软的背景墙，微妙而复杂的印花地毯给空间增添了厚重感。漂亮的亚麻抱枕软化了整个空间的生硬感，让卧室在宁静中又不失豪华气质。浴室内，大理石质地的地板和墙壁，将房间内庄重严肃的质感延伸开来。

紧邻主卧的儿童房的小主人是一个上高中的男孩。空间采用了驼色与深绿色，将现代和古典元素融入其中，地毯采用了充满活力的图案及具有欧式元素的软装点缀，为房间里增添了几分俏皮与活泼。

绅士简约主义
The Simplicity of Gentleman

项目名称：中航城 B2-2
设计单位：SCD（香港）郑树芬设计事务所
设计师：郑树芬
项目地点：贵州省贵阳市

Project Name: China Aviation City B2-2
Design Company: SCD (HongKong) Simon Chong Design Consultants Limited
Designer: Simon Chong
Project Location: Guiyang City , Guizhou Province

一种彬彬有礼的气质，蔓延在整个空间中。简练、硬朗、高贵，精致到一丝不苟。想象中，可以感受到沉稳内敛的空间精神……

设计师将简约与复古相结合的空间情调贯穿始终。在软装搭配上，选用了具有麻面质感的材料，与壁纸的肌理纹样及地毯的条纹图案相搭配，创造出素雅、简朴的肌理感。

A polite temperament, spread throughout the space. Tough, noble, delicate to be strict in one's demands. Imagine, you can feel the spirit of calm and reserved space......

Designers will be simple and retro space, the combination of the. In soft outfit collocation, selected with a pitted surface texture of the material, and wallpaper texture patterns and carpet fringe pattern match to create elegant, simple texture.

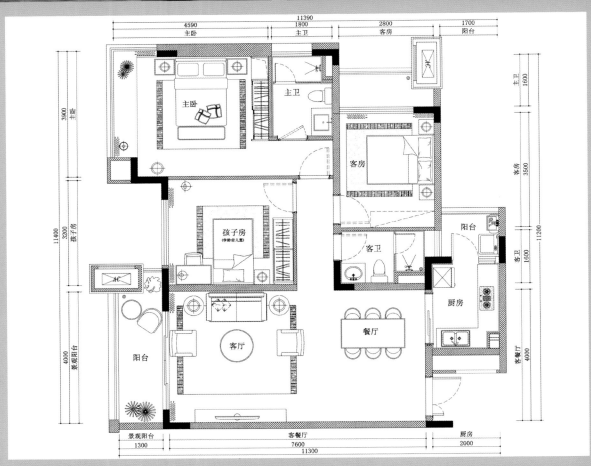

平面布置图

漂亮时光

爱的守护
Guardian of Love

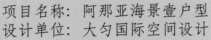

项目名称：阿那亚海景壹户型
设计单位：大勺国际空间设计
设计师：林宪政
软装设计：上海太舍馆贸易有限公司
项目面积：115 ㎡
开发商：秦皇岛天行九州旅游置业开发有限公司

Project Name: Anaya Seaview Room –Type1
Design Company: SD SYMMETRY International Space Design
Designer: Lin Xianzheng
Soft Decoration: Shanghai Mogadeco Trade Co., Ltd.
Project Area: 115 ㎡
Developer: Qinhuangdao Tianxing Jiuzhou Tourism Real Estate Development Co. Ltd.

"我有一所房子，面朝大海，春暖花开"，海子的诗歌描写出很多人梦想中的生活。

开发者将自己定义为美好生活的服务商，依托这片大海，发掘它带给人的快乐，营造、完善、深化这片土地的气质。

其实现代人都需要一个躲起来的空间，不在于这个空间的大小。现代人很少有机会真正独处，大部分时间我们跟其他人一起，无论是跟工作伙伴还是家人，都是群居的，那么他就需要一个独处的空间，这个空间是他自己看自己的空间。

设计师希望提供这样的空间，里面摆什么东西不重要，重要的是在这个空间，他可以得到心灵的寄托。

从做设计的角度，我觉得应该有一种"中间"的可能，这个空间不需要那么漂亮，不用贴金覆银，像一个车库就好，不用太精致，也不怕弄坏掉，在那里可能会放置很多与个人记忆有关的收藏品。但是，这个空间至少是漂亮的。就像女生有很多漂亮的衣服，就应该摆在一个漂亮的地方，如果这个空间不够漂亮，她不会觉得自己有多么重要。

这就是看与被看的关系。无论是纪念性还是仪式感，它肯定有自己的设计方式，比如线条、挑高、光线，但它最终肯定与个人的记忆有关。我希望客户可以把自己的心放进去。

"I have a house, facing the sea, spring", the poems describe many people dream of life.

Developers will define themselves as a good life service providers, relying on this sea, to explore it brings people happiness, to create, improve and deepen the land of the temperament.

In fact, modern people need a hiding space, not the size of the space. Modern people rarely have the opportunity to truly be alone, and most of the time we are living together with other people, whether it is living with a partner or family, and he needs a space to be alone.

Designers hope to provide such a space, which is not important to put what is important in this space, he can get the sustenance of the heart.

From the design point of view, I think there should be a kind of "intermediate", the space does not need to be so beautiful, not gold and silver, like a garage is good, not too fine, also is not afraid of broken off, there may will be placed a lot of personal memory related Collectibles. But this space is at least pretty. Like girls have many beautiful clothes, they should be placed in a very beautiful place, if this space is not beautiful, she will not feel how important.

This is to see the relationship with the. Whether it is a memorial ceremony, it is sure to have their own design, such as lines, high, light spilled in, but it is ultimately related to the memory of the individual. I hope that customers can put their hearts in.

定义奢华新概念
Define the New Concept of Luxury

设计公司：重庆品辰装饰工程设计有限公司
空间设计师：庞一飞、李健
软装设计师：夏婷婷、黄琳
项目地点：云南省昆明市
项目面积：220 ㎡
主要材料：珍珠鱼皮、钢琴烤漆、壁纸、贝母马赛克、琉璃灯、钛金不锈钢、石材

Design Company: Chongqing Pinchen Decoration Engineering Design Co., Ltd.
Space Designers: Pang Yifei , Li Jian
Soft Decoration Designers: Xia Tingting , Huang Lin
Project Area: 220 ㎡
Project Location: Kunming City , Yunnan Province
Main Materials: Pearl Skin, Piano Paint , Wallpaper, Fritillary Mosaic ,Glass Lamps, Stainless Steel Titanium, Stone

豪宅软装十风格·复古雅皮

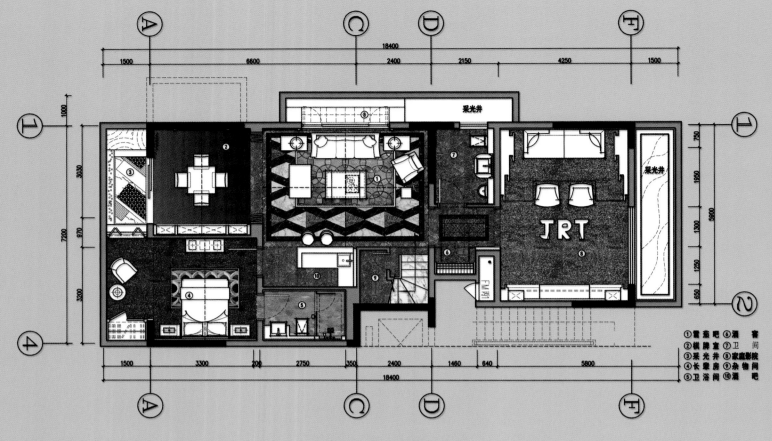

负一层平面布置图

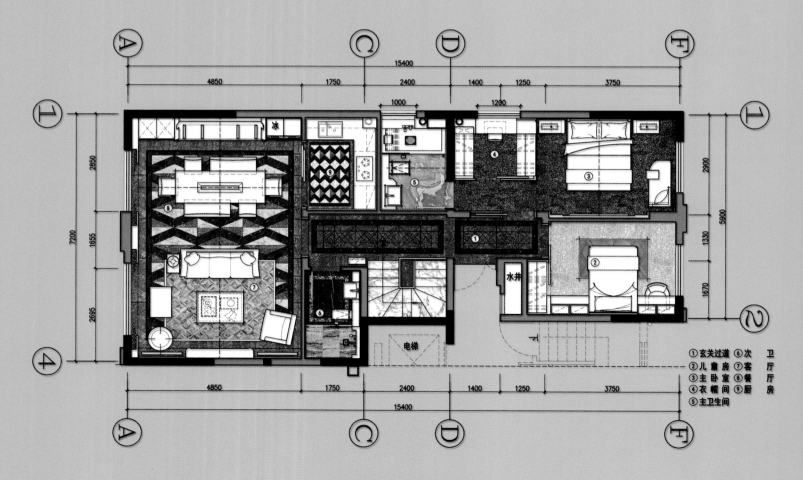

底跃平面布置图

豪宅软装十风格·复古雅皮

品辰拒绝无趣的装修，最大程度地做到守住这静谧的时光，打造出与现代生活相契合的空间。

在前期的设计调研中，品辰设计小组前往项目所在地考察，使其在设计细节的把控上更加仔细，营造出迷人的氛围。无论是有机切割的拼花地板还是灯光设计，空间沉溺于几何之美，又蕴含着五行八卦的原理。

采光井的增建，让人在室内可以自由地呼吸大自然的新鲜。这是一种顺畅的自然空间的过渡，还有可以看见的四季春光。

餐厅朦胧的琉璃灯在轻轻吟诵着浪漫，珍珠鱼皮演绎着尊贵。

本案结合现代视角与当地传统设计，展现出品辰设计对于人类情感的理解及人性化体验的关注。

Pinchen Design refused to do the decoration, the greatest degree of doing to keep this quiet time, to create a space to adapt to the modern life.

In the preliminary design research, product design team to go to the magic all inspection, so that the design details of the control more carefully, create a charming atmosphere. Both the organic cutting parquet or lighting design, indulge in the beauty of geometry, but also contains five gossip.

Additional lighting wells, make indoor air to breathe freely the fresh nature. A smooth and natural transition, controllable seasons of spring.

The restaurant in the dim lamp glass gently singing a romantic interpretation of the noble, Pearl skin.

With the modern perspective and the local traditional design, it also shows that the product design is concerned about human emotion and human nature.

图书在版编目(CIP)数据

漂亮时光·豪宅软装十风格. 上 / 海燕编. —武汉：华中科技大学出版社，2016.1
ISBN 978-7-5609-8999-0

Ⅰ. ①漂… Ⅱ. ①海… Ⅲ. ①住宅－室内装饰设计 Ⅳ. ①TU241

中国版本图书馆CIP数据核字(2015)第218027号

漂亮时光·豪宅软装十风格（上）

海燕 编

出版发行：华中科技大学出版社（中国·武汉）
地　　址：武汉市武昌珞喻路1037号（邮编：430074）
出 版 人：阮海洪

责任编辑：赵爱华	责任监印：秦　英
责任校对：胡　雪	装帧设计：李　乐

印　　刷：深圳当纳利印刷有限公司
开　　本：965 mm×1270 mm　1/16
印　　张：20
字　　数：288千字
版　　次：2016年1月第1版第1次印刷
定　　价：338.00元（USD 69.99）

投稿热线：(010)64155588-8000
本书若有印装质量问题，请向出版社营销中心调换
全国免费服务热线：400-6679-118　竭诚为您服务
版权所有　侵权必究